Electrophysiology

The Practical Approach Series

SERIES EDITORS

D. RICKWOOD
Department of Biology, University of Essex
Wivenhoe Park, Colchester, Essex CO4 3SQ, UK

B. D. HAMES
Department of Biochemistry and Molecular Biology, University of Leeds
Leeds LS2 9JT, UK

Affinity Chromatography
Anaerobic Microbiology
Animal Cell Culture (2nd Edition)
Animal Virus Pathogenesis
Antibodies I and II
Biochemical Toxicology
Biological Membranes
Biomechanics—Materials
Biomechanics—Structures and Systems
Biosensors
Carbohydrate Analysis
Cell–Cell Interactions
Cell Growth and Division
Cellular Calcium
Cellular Neurobiology
Centrifugation (2nd Edition)
Clinical Immunology
Computers in Microbiology
Crystallization of Nucleic Acids and Proteins
Cytokines
The Cytoskeleton
Diagnostic Molecular Pathology I and II
Directed Mutagenesis
DNA Cloning I, II, and III
Drosophila
Electron Microscopy in Biology
Electron Microscopy in Molecular Biology
Electrophysiology
Enzyme Assays
Essential Molecular Biology I and II
Experimental Neuroanatomy

Fermentation
Flow Cytometry
Gel Electrophoresis of Nucleic Acids (2nd Edition)
Gel Electrophoresis of Proteins (2nd Edition)
Genome Analysis
Haemopoiesis
HPLC of Macromolecules
HPLC of Small Molecules
Human Cytogenetics I and II (2nd Edition)
Human Genetic Diseases
Immobilised Cells and Enzymes
In Situ Hybridization
Iodinated Density Gradient Media
Light Microscopy in Biology
Lipid Analysis
Lipid Modification of Proteins
Lipoprotein Analysis
Liposomes
Lymphocytes
Mammalian Cell Biotechnology
Mammalian Development
Medical Bacteriology
Medical Mycology
Microcomputers in Biochemistry
Microcomputers in Biology
Microcomputers in Physiology
Mitochondria
Molecular Genetic Analysis of Populations
Molecular Neurobiology
Molecular Plant Pathology I and II
Monitoring Neuronal Activity
Mutagenicity Testing
Neural Transplantation
Neurochemistry
Neuronal Cell Lines
Nucleic Acid and Protein Sequence Analysis
Nucleic Acid Hybridisation
Nucleic Acids Sequencing
Oligonucleotides and Analogues
Oligonucleotide Synthesis
PCR

Peptide Hormone Action

Peptide Hormone Secretion

Photosynthesis: Energy Transduction

Plant Cell Culture

Plant Molecular Biology

Plasmids

Pollination Ecology

Post-implantation Mammalian Embryos

Preparative Centrifugation

Prostaglandins and Related Substances

Protein Architecture

Protein Engineering

Protein Function

Protein Purification Applications

Protein Purification Methods

Protein Sequencing

Protein Structure

Protein Targeting

Proteolytic Enzymes

Radioisotopes in Biology

Receptor Biochemistry

Receptor–Effector Coupling

Receptor–Ligand Interactions

Ribosomes and Protein Synthesis

Signal Transduction

Solid Phase Peptide Synthesis

Spectrophotometry and Spectrofluorimetry

Steroid Hormones

Teratocarcinomas and Embryonic Stem Cells

Transcription and Translation

Virology

Yeast

Electrophysiology
A Practical Approach

Edited by
D. I. WALLIS
*Department of Physiology,
University of Wales College of Cardiff,
Cardiff*

—at—
OXFORD UNIVERSITY PRESS
Oxford New York Tokyo

Oxford University Press, Walton Street, Oxford OX2 6DP
Oxford New York Toronto
Delhi Bombay Calcutta Madras Karachi
Kuala Lumpur Singapore Hong Kong Tokyo
Nairobi Dar es Salaam Cape Town
Melbourne Auckland Madrid
and associated companies in
Berlin Ibadan

Oxford is a trade mark of Oxford University Press

A Practical Approach 🛑 is a registered trade mark
of the Chancellor, Masters, and Scholars of the University of Oxford
trading as Oxford University Press

Published in the United States
by Oxford University Press Inc., New York

© Oxford University Press, 1993

All rights reserved. No part of this publication may be
reproduced, stored in a retrieval system, or transmitted, in any
form or by any means, without the prior permission in writing of Oxford
University Press. Within the UK, exceptions are allowed in respect of any
fair dealing for the purpose of research or private study, or criticism or
review, as permitted under the Copyright, Designs and Patents Act, 1988, or
in the case of reprographic reproduction in accordance with the terms of
licences issued by the Copyright Licensing Agency. Enquiries concerning
reproduction outside those terms and in other countries should be sent to
the Rights Department, Oxford University Press, at the address above.

This book is sold subject to the condition that it shall not,
by way of trade or otherwise, be lent, re-sold, hired out or otherwise
circulated without the publisher's prior consent in any form of binding
or cover other than that in which it is published and without a similar
condition including this condition being imposed
on the subsequent purchaser.

Users of books in the Practical Approach series are advised that prudent
laboratory safety procedures should be followed at all times. Oxford
University Press makes no representation, express or implied, in respect of
the accuracy of the material set forth in books in this series and cannot
accept any legal responsibility or liability for errors or omissions
that may be made.

A catalogue record for this book is available from the British Library

Library of Congress Cataloging in Publication Data
Electrophysiology: a practical approach/edited by D. I. Wallis.—
1st ed.
(Practical appoach series)
1. Electrophysiology. 2. Molecular neurobiology. I. Wallis, D. I.
II. Series.
[DNLM: 1. Cytological Techniques. 2. Histological Techniques.
3. Electrophysiology—methods—laboratory manuals. 4. Nerve Tissue—
physiology—laboratory manuals. 5. Neural Transmission—
physiology—laboratory manuals. 6. Neurons—drug effects—
laboratory manuals. 7. Neurons—physiology—laboratory manuals.
WL 25 E38]
QP341.E344 1993 599'.0188—dc20 92-49953
ISBN 0–19–963347–9 (pbk.)
ISBN 0–19—963348–7 (hbk.)

Typeset by Cambrian Typesetters, Frimley, Surrey
Printed by Information Press Ltd., Eynsham, Oxon.

Preface

Researchers often wish to extend the range of techniques at their command. Biologists in particular may require an introduction to electrophysiological methods. This book is intended to provide a very practical and specific introduction to certain techniques important in the study of single cells and of complex nervous tissue. The emphasis is on procedures and step-by-step methodology.

The book has four sections. The first, on membrane currents, includes chapters on peripheral neurones, lens epithelial cells, sinoatrial node cells, and on the *Xenopus* oocyte as an expression system in molecular biology. Section 2 deals with synaptic transmission and considers the pharmacological analysis of transmission in brain slices, the quantal analysis of synaptic transmission, and the intracellular staining of enteric neurones. Section 3, under the heading central nervous system, includes a description of whole-cell patch-clamping of cells in spinal cord slices and, more generally, the use of the spinal cord as an *in vitro* preparation. This section also includes an approach towards a mathematical description of neuronal firing patterns. In the final section, nerve cell pharmacology, consideration is given to the use of microiontophoresis in the central nervous system and to electrophysiological techniques for studying the quantitative pharmacology of the isolated spinal cord.

Although I have taken the view that electrophysiology, as an approach to research, probably finds its greatest application in the area of neurobiology, I have attempted as broad an approach as possible within the confines of the format. Nevertheless, consideration of the techniques of measurement and analysis of field potential data such as EEG, EMG, ECG, and ERG seemed beyond the scope of this volume, even though such electrophysiological techniques may be widely used as research tools. Further, I have tried to avoid too much overlap with some other excellent texts. To some extent I hope this book fills a niche between *Microelectrode techniques: the Plymouth workshop handbook* edited by N. B. Standen, P. T. A. Gray, and M. J. Whitaker (1987), *Single channel recording* edited by B. Sakmann and E. Neher (1983), and the volume in this series, *Monitoring neuronal activity: a practical approach* edited by J. A. Stamford. Another companion volume in this series (*Receptor–effector coupling: a practical approach*) has an excellent chapter on patch-clamping, entitled 'Molecular pharmacology of ion channels using the patch-clamp'.

In addition to the main chapters, this book contains a short appendix which lists the names and addresses of suppliers mentioned in the text.

It is hoped that the book will be of substantial use to research students and research scientists, and may also be of value to undergraduates in their final

Preface

year who undertake a research project. My thanks are due to all the authors for their fine contributions and for their understanding and perseverance in preparing them.

Cardiff
April 1992

D. I. WALLIS

Contents

List of contributors   xix

Abbreviations   xxi

PART 1 MEMBRANE CURRENTS

1. Recording membrane currents of peripheral neurones in short-term culture   3
Catherine Stansfeld and Alistair Mathie

1. Introduction	3
2. Preparation of cells	4
Is a sterile technique needed for a short-term preparation?	
Do I need a tissue culture laboratory?	4
Preparation of reagents and substrates	4
Finding the SCG and nodose ganglia	5
Culture of ganglion cells	8
Critical observation of cells	10
Control of cell adhesion and neurite outgrowth	10
3. Recording from ganglion cells	12
Choice of voltage-clamp amplifier	12
How well can these cells be clamped?	13
Leak—and how to deal with it	15
4. Solution design and adequacy of exchange between pipette and cell interior	17
General points	17
Time course of cell dialysis	18
Rundown of current using whole-cell and single channel recording	20
Other considerations	21
5. Channels and receptors	22
Calcium currents	22
Potassium currents	25
Receptors	27
Summary	27
Acknowledgements	27
References	28

Contents

2. Transport mechanisms in ocular epithelia 29

T. J. C. Jacob, J. W. Stelling, Amanda Gooch, and Jin Jun Zhang

1. Introduction 29
2. Lens epithelium 30
    The 'double chamber' applied to the lens epithelium 30
    Verification of measurements 32
    Resistance measurements 34
3. Ciliary epithelial cells 35
    Preparation of fresh cells (pigmented and non-pigmented) 35
    Preparation of cells for culture 37
    Preparation of electrodes (pipettes) 39
4. Recording from epithelial cells using the patch-clamp 41
    Patch recording 41
    Whole-cell recording 44
    Data analysis 46
5. Differences between fresh and cultured cells 46
    Size difference 46
    Disappearance of inward rectifier 47

References 47

3. Isolation of sinoatrial node cells 49

H. F. Brown

1. Introduction 49
2. Separation of sinoatrial node cells 49
    General introduction 49
    Appearance of healthy SA node cells 49
    Methods of isolating SA node cells 51
3. The shape of sinoatrial node cells: spindles, spheres, or spiders? 55
4. Electrophysiological study of the isolated sinoatrial node cell 56
    Internal solution for the patch pipette 56
    Whole-cell patch electrode recording from the isolated SA node cell 57

Contents

5. Membrane currents	58
Time-dependent currents	58
Time-independent currents	60
Run-down of membrane currents	61
6. Permeabilized patch technique	61
Acknowledgements	62
References	62

4. Electrophysiology of *Xenopus* oocytes: an expression system in molecular neurobiology   65

S. P. Fraser, C. Moon, and M. B. A. Djamgoz

1. Introduction	65
2. Endogenous characteristics	65
Morphological properties	66
Electrophysiological properties	66
Activation of second-messenger systems and G-proteins	66
3. Strategies for expression studies	69
Preparation of mRNA	69
Surgery of *Xenopus* and removal of oocytes	70
Preparation of single oocytes	72
4. Injection of mRNA	73
Micropipettes	73
Injection apparatus	73
Injection procedure	74
Maintenance of oocytes	75
5. Electrophysiological recordings	75
Two-electrode voltage-clamp	76
Patch-clamp	76
Perfusion system	78
6. Complementary screening techniques	79
7. Recent expression studies	80
Expression and characterization of specific subunits	80
Co-translation of mRNAs from different sources	81
Translation of mRNAs following site-directed mutagenesis	81
Expression of a receptor that inhibits a second messenger	82
Expression of neurotransmitter receptors from invertebrates	83
References	85

PART II SYNAPTIC TRANSMISSION

5. Pharmacological analysis of synaptic transmission in brain slices   89
Graeme Henderson

    1. Introduction   89
    2. Electrical recording from brain slices   90
        Preparation of brain slices   90
        Recording chamber   92
        Bathing solution   92
        Methods for evoking and recording post-synaptic responses   93
    3. Analysis of postsynaptic potentials   97
        Blockade of synaptic transmission   97
        Pharmacological dissection of multi-component postsynaptic potentials   98
        Ionic mechanism   103
        Sites of action of drugs modifying synaptic transmission   105
    4. Conclusions   107
        References   107

6. Quantal analysis of synaptic transmission   109
Bruce Walmsley

    1. Introduction   109
    2. Why use quantal models?   111
    3. Structural basis and interpretation of quantal models   111
        What is a release site?   111
        What is a quantum?   113
        What is N?   115
    4. Recording and measurement of synaptic potentials and currents   116
        Synaptic potentials or currents?   116
        Automatic baseline correction   117
        Initial analysis of evoked responses and noise contamination: choosing baseline and measurement regions   118
        Spontaneous potentials and currents   122

Contents

5. Models of amplitude fluctuations of synaptic potentials (currents)	123
Maximum likelihood method	124
Noise contamination and deconvolution	126
Formulating a model	127
Discrete PDF models	128
Quantal variability and spontaneous potentials (currents)	136
Other models and procedures	138
Acknowledgements	139
References	139

7. Intracellular staining of enteric neurones   143

Gordon M. Lees

1. Introduction	143
2. Equipment required	143
3. Solutions, materials, and dyes	145
Krebs' solution	145
Other solutions required	146
Other materials	146
4. Procedures	146
Preparation of myenteric plexus	146
Preparation of submucous plexuses	148
Histochemical detection of myenteric and submucous plexuses	149
Fixation and processing of specimens	150
5. Ancillary techniques	151
Fluorescence microscopy	151
Fluorescence immunohistochemistry	151
Photomicrography	152
6. Intracellular staining	152
Choice of intracellular markers	152
Requirements and precautions	154
Quality control	155
Procedures for intracellular staining	156
Cell types most likely to be stained	156
Problems	156
7. Intracellular electrophysiological recording	158
Hints on good impalement with fluorescent dye-filled microelectrodes	158
Properties of enteric neurones	159
Pitfalls in recording and neuronal classification	160

Contents

Acknowledgements	163
References	163

PART III CENTRAL NERVOUS SYSTEM

8. Whole-cell patch-clamp recording from neurones in spinal cord slices 169

Tony E. Pickering, Dave Spanswick, and Steve D. Logan

1. Introduction	169
2. Spinal cord slice preparation	171
3. Maintenance of slices	173
4. Whole-cell recording from neurones in slices	174
The rig	174
Patch pipettes	176
Pipette solutions	177
Obtaining whole-cell recordings	179
Perforated-patch recordings	183
5. Visualization of neurones	183
6. Results	185
7. Prospects	185
Acknowledgements	187
References	187

9. The spinal cord as an *in vitro* preparation 189

Jeffery Bagust

1. Introduction	189
2. Species and age	190
3. Dissection	191
Equipment	191
Anaesthetic	193
Removal of vertebral column	194
Dissection of the cord	195
Hemisection	199
Stripping	200

Contents

4. The perfusion system	201
Circulation	201
The recording chamber	202
Flow rate	202
Cleaning	203
5. Recording	203
Mounting and recovery	203
Recording techniques	206
Stimulation	208
Example recordings/manipulations	208
6. Temperature	210
7. Troubleshooting	211
Possible problems and remedies	211
If the preparation is completely dead	213
References	213

10. Mathematical description of neuronal firing patterns   215

J. L. Hindmarsh and R. M. Rose

1. Introduction	215
2. Theory	215
Prediction of the time course of the membrane potential	215
The FitzHugh model	216
The 'state space' and 'state diagram'	218
The 'tangent vector'	219
The 'nullclines'	220
The 'equilibrium points'	221
The 'limit cycle'	222
The effect of varying a parameter	222
The effect of an external current step	223
Differential equation models	224
Choice of step size	225
3. The mathematical models	226
The repetitively firing neurone	226
Triggered firing	227
Adaptation	229
Burst generation	232
4. Conclusions	234
References	235
Appendix	235

PART IV NERVE CELL PHARMACOLOGY

11. Iontophoresis in the mammalian central nervous system   239
M. H. T. Roberts and T. Gould

   1. Introduction   239
      Basic principles   239
   2. Preparation of micropipettes   240
      Manufacture of blanks   240
      Pulling micropipettes   242
      Preparing the electrodes for filling   244
      Drug solutions and electrode filling   244
   3. The release of drugs from micropipettes   247
      Hydrostatic efflux and diffusion   248
      Electro-osmosis   248
      Microiontophoresis   248
   4. Experimental measurement of drug release   249
      Diffusional release   249
      pH and the effects of competing ions in the solutions   249
      Retaining currents   250
      Bradshaw's model of ion movements in a pipette   254
   5. Practical experimental design   254
      A recommended method—quantification of drug release   256
      Determination of comparative potency in biological experiments   259
      The control of extracellular current artefacts   259
      Changes in amplitude of action potentials   260
      Signal-to-noise ratio   261
      The electronic timing of drug application   261

   References   263

12. Electrophysiological techniques for studying the quantitative pharmacology of the isolated spinal cord   265
D. I. Wallis

   1. Introduction   265

Contents

2. Extracellular recording of population responses — 265
  Baths — 267
  Preparation of cord — 267
  Extracellular methods of recording — 268
  Significance of uptake mechanisms in the cord — 272
  Ipsilateral and contralateral reflexes — 273
  Tail-attached spinal cord preparations — 279
  Other uses of the hemisected cord technique — 280

3. Intracellular recording from hemisected cord — 280
  Stimulation of the dorsal root and the cord surface — 280
  Application of transmitter chemicals — 282
  Muscle-attached preparations — 282

Acknowledgements — 284

References — 284

Appendix

A1 Addresses of some suppliers — 285

Index — 289

Contributors

JEFFERY BAGUST
Department of Physiology and Pharmacology, University of Southampton, Southampton SO9 3TU, UK

H. F. BROWN
University Laboratory of Physiology, Parks Road, Oxford OX1 3PT, UK

M. B. A. DJAMGOZ
Department of Biology, Imperial College, London SW7 2BB, UK

S. P. FRASER
Department of Biology, Imperial College, London SW7 2BB, UK

AMANDA GOOCH
Department of Physiology, University of Wales College of Cardiff, Cardiff CF1 1SS, UK

T. GOULD
Department of Physiology, University of Wales College of Wales, Cardiff CF1 1SS, UK

GRAEME HENDERSON
Department of Pharmacology, University of Bristol, University Walk, Bristol BS8 1TD, UK

J. L. HINDMARSH
Department of Mathematics, University of Wales College of Cardiff, Cardiff CF1 1SS, UK

T. J. C. JACOB
Department of Physiology, University of Wales College of Cardiff, Cardiff CF1 1SS, UK

GORDON M. LEES
Department of Biomedical Sciences, Marischal College, Aberdeen University, Aberdeen AB9 1AS, UK

STEVE D. LOGAN
Department of Physiology, University of Birmingham, Birmingham B15 2TT, UK

ALISTAIR MATHIE
Department of Pharmacology, Royal Free Hospital School of Medicine, Rowland Hill Street, London NW3 2PF, UK

Contributors

C. MOON
Department of Biology, Imperial College, London SW7 2BB, UK

TONY E. PICKERING
Department of Physiology, University of Birmingham, Birmingham B15 2TT, UK

M. H. T. ROBERTS
Department of Physiology, University of Wales College of Cardiff, Cardiff CF1 1SS, UK

R. M. ROSE
Department of Physiology, University of Wales College of Cardiff, Cardiff CF1 1SS, UK

DAVE SPANSWICK
Department of Physiology, University of Birmingham, Birmingham B15 2TT, UK

CATHERINE STANSFELD
Department of Pharmacology, University College London, Gower Street, London WC1E 6BT, UK

J. W. STELLING
Department of Physiology, University of Wales College of Cardiff, Cardiff CF1 1SS, UK

D. I. WALLIS
Department of Physiology, University of Wales College of Cardiff, Cardiff CF1 1SS, UK

BRUCE WALMSLEY
The Neuroscience Group, University of Newcastle, University Drive, Callaghan, NSW 2308, Australia

JIN JUN ZHANG
Department of Physiology, University of Wales College of Cardiff, Cardiff CF1 1SS, UK

Abbreviations

ACh	acetylcholine
AMPA	α-amino-3-hydroxy-5-methyl-4-isoxazolepropionic acid
AlF_4^-	aluminium fluoride complex
α-BUTX	alpha-bungarotoxin
AP4	L-2-amino-4-phosphonobutyric acid
APV	D-2-amino-5-phosphonopentanoic acid
AHP	after-hyperpolarization
BAPTA	1,2-bis(O-aminophenoxy)ethane-N,N,N',N'-tetra acetic acid
pCA	para-chloroamphetamine
C_M	membrane capacitance
CMF	calcium- and magnesium-free solution
CNQX	6-cyano-7-nitroquinoxaline-2,3-dione
CNS	central nervous system
CON FAST	contralateral fast response
CON SLOW	contralateral slow response
CPP	3-((±)-2-carboxypiperazin-4-yl) propyl-1-phosphonic acid
CR	concentration–response
aCSF	artificial cerebrospinal fluid
DAG	diacylglycerol
DEPC	diethyl pyrocarbonate
DHP	dihydropyridine
DMSO	dimethylsulphoxide
DPBS	Dulbecco's phosphate-buffered saline
DP-5-CT	dipropyl-5-carboxamidotryptamine
DRG	dorsal root ganglion
EGTA	ethyleneglycol bis-(β-aminoethyl)ether-N,N,N',N'-tetra acetic acid
E–M	expectation–maximization
ENS	enteric nervous system
EP	equilibrium point
EPSP	excitatory postsynaptic potential
E_{rev}	reversal potential
FCS	fetal calf serum
FITC	fluorescein isothiocyanate
GABA	gamma-aminobutyric acid
Hepes	N-2-hydroxyethylpiperazine N'-2-ethanesulphonic acid
HRP	horseradish peroxidase
5-HT	5-hydroxytryptamine
I_A	A current
$I_{b,Na}$	inward background current
I_{DR}	delayed rectifier current (also I_K)
I_f	hyperpolarization-activated inward current
$I_{K(AHP)}$	calcium-activated potassium current associated with after-hyperpolarization

Abbreviations

$I_{K(Ca)}$	calcium-activated potassium current
I_K	delayed rectifier current (also I_{DR})
IBMX	3-isobutyl-1-methylxanthine
IHC	immunohistochemistry
IPSI SLOW	ipsilateral slow response
IPSP	inhibitory postsynaptic potential
IP_3	inositol triphosphate
MLE	maximum likelihood estimator
MSR	monosynaptic reflex
NBQX	6-nitro-7-sulphamoyl-benzo(f)quinoxaline-2,3-dione
NBT–NADH	nitro-blue tetrazolium-reduced nicotinamide dinucleotide solution
NGF	nerve growth factor
NMDA	N-methyl-D-aspartic acid
PBS	phosphate-buffered saline
PDF	probability density function
PSR	polysynaptic reflex
PT	pars tuberalis
Q	quantal size
R_L	leak resistance
R_M	membrane resistance
RNase	ribonuclease
R_p	summed access and membrane resistance
R_S	series resistance
R_{TE}	transepithelial resistance
SA	sinoatrial
SCG	superior cervical ganglion
SEVC	single electrode voltage-clamp
SPN	sympathetic preganglionic neurones
Tris	tris(hydroxy methyl) amino methane
TTX	tetrodotoxin
VIP	vasoactive intestinal peptide
WCR	whole-cell recording

PART I

Membrane currents

1

Recording membrane currents of peripheral neurones in short-term culture

CATHERINE STANSFELD and ALISTAIR MATHIE

1. Introduction

In comparison with central neurones, ganglion cells are attractive because they are accessible collections of relatively uniform cell types, easily identified visually. An added advantage is that they are generally large—dorsal root ganglion (DRG) cells being 10–60 µm and sympathetic superior cervical ganglion (SCG) cells being 15–40 µm—thus, they are readily accessible to electrophysiological techniques. Quite apart from questions concerning the physiology of the sensory system, the unipolar DRG and nodose ganglion cells can be viewed as model spherical neurones to which can be applied more general questions concerning pre- or post-synaptic membrane processes. They have, for example, been used extensively to examine biochemical processes governing calcium channel function. Thus, they are an invaluable electrophysiological tool from which we can gain insights into the mechanisms at work in less accessible neuronal membranes. On the other hand, SCG cells are normally multipolar, and so lack the ideal compactness afforded by the sensory ganglion cells; however, they temporarily adopt this ideal spherical form when dispersed from the ganglion. At this stage their membrane currents can be studied without the complications of dendrites. In this chapter, we concentrate on a short-term preparation of isolated cells—to be studied within 24 h—and have chosen sympathetic SCG cells as our example. The purpose of this preparation is to achieve rapid cell dissociation for electrophysiology in which high cell yield is not important (4000 cells per day is usually enough for most electrophysiologists). Our emphasis is on how to optimize conditions for good control of the membrane voltage, and how to make use of this in interpreting the membrane currents seen.

This chapter is intended to be complementary to several other highly relevant chapters in this series, with which we have attempted to avoid too great an overlap. For example Lindsay and colleagues (1) also provide a methodology for culture of peripheral ganglion cells and can be consulted for more extensive coverage. Finkel (2) covers details of voltage-clamp methodology; Dempster (3), methods of data acquisition, storage, and analysis; Gurney (4), the art of patch-clamping; Kostyuk (5), intracellular perfusion of ganglion cells and a description of some of their ionic currents; Lancaster (6), isolation of potassium currents; and Borg-Graham (7), non-linear properties of excitable membranes. See also Chapters 2, 3, and 8 in this volume.

2. Preparation of cells

2.1 Is a sterile technique needed for a short-term preparation? Do I need a tissue culture laboratory?

The short answer to the first question is probably 'yes' and to the second one, 'no'. The isolated cells are to be kept healthy for up to 24 h. Both cells and their culture medium will be targets for bacteria, yeasts, fungus, and virus which will multiply rapidly. Culture dishes are designed to have free access of air under their lids, and the incubators in which they are maintained ensure a continuous air circulation over the dishes. Airborne sources of infection readily transfer to both the incubator and its contents. The incubator is not itself a sterile environment but it is regarded as 'clean', and it should not become an infective source. A short-term user must avoid endangering the condition of other longer term cultures sharing the same incubator. In practice, this means use of protocols which minimize transfer of infective organisms into the culture environment and which minimize the conditions in which they thrive.

An additional point is that the culture media used during cell adhesion are both expensive and take time to prepare. Normally, sufficient solutions are made up for several cultures and their sterility is important.

However, short-term culture does not necessitate an elaborately equipped tissue culture room with laminar flow hood. A clean corner of a laboratory away from general traffic and draughts, some kind of enclosed cabinet within which the more crucial sterile procedures may be carried out, a small centrifuge, and an incubator are the principal requirements. A fibre optic light source, a dissection microscope and a good phase-contrast microscope for monitoring the cells' progress, fridge and freezer space for the tissue culture reagents, and an autoclave somewhere in the building completes the needs.

2.2 Preparation of reagents and substrates

2.2.1 Preparation of substrate

We use laminin (Sigma), collagen (Type I, Vitrogen 100, Imperial Laboratories), or poly-D-lysine (mol. wt 300 000, Sigma).

For laminin:

(a) Prepare at 10 µg/ml in buffered Hanks' solution.
(b) Place a 200 µl blob in the centre of each 35 mm culture dish at least 2 h before use.
(c) Rinse before plating.

For collagen:

(a) Spread 200 µl of stock solution (approximately 3 mg/ml in 0.012 M HCl) in the centre of each 35 mm culture dish.
(b) Leave overnight to dry.
(c) Rehydrate for 20 min prior to use, and rinse with buffer to neutralize acidity.

For polylysine:

(a) Treat culture dish with 100 µg/ml in H_2O for a minimum of 2 h.
(b) Wash with growth medium immediately before plating cells.

2.2.2 Solutions and culture media

The basic solutions and culture media which we generally use are:

(a) L_{15} (Liebovitz) (Gibco)
(b) Hanks' Balanced Salts (calcium- and magnesium-free) (Gibco) with 10 mM Hepes added
(c) enzymes: collagenase 375 IU/ml (Worthington Biochemicals) and trypsin 1 mg/ml (Sigma XII-S) made up in the Hanks' Balanced Salts, calcium- and magnesium-free with added Hepes, as above. Each enzyme is prepared separately immediately before use, together with 6 mg/ml bovine serum albumin (Sigma)
(d) growth medium, consisting of:

   i. L_{15} (Liebovitz) (Gibco)
   ii. glutamine 2 mM (Gibco)
   iii. fetal calf serum 10% (Gibco)
   iv. sodium bicarbonate 0.21% (Gibco)
   v. mouse nerve growth factor (NGF) 7S 10 ng/ml (Sigma)
   vi. D-glucose 0.69%
   vii. Pen/Strep: 50 IU penicillin, 50 µg streptomycin/ml (Gibco)

2.3 Finding the SCG and nodose ganglia

The SCG and nodose ganglia lie together bilaterally located in the cervical region, close to the carotid artery bifurcation which forms the internal and

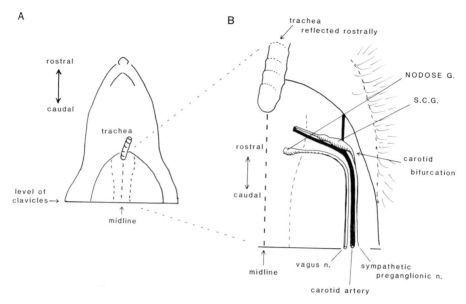

Figure 1. Dissection of superior cervical ganglion (S.C.G.) and nodose ganglion (NODOSE G.) in the rat. (A) Diagrammatic view of ventral surface of neck and head of rat. The skin is reflected away from the neck region and the cut end of the trachea has been turned rostrally. (B) Expanded view of one half of the dissection field showing the position of the ganglia alongside the carotid artery. n., nerve.

external carotid arteries. The dissection of either of them is virtually identical and is easy and quick, but four ganglia are needed for a culture, i.e. two animals. The nodose, though frequently termed a 'sympathetic' ganglion, is in fact purely primary afferent: it is the cranial equivalent of a dorsal root ganglion, except that more than 90% of the neurones are sensory C-fibre neurones. Refer to *Figure 1* whilst following *Protocol 1*.

Protocol 1. Finding the SCG and nodose ganglia

1. Kill a 17-day-old rat by placing it in a container into which 100% CO_2 is introduced, then decapitate, cutting close to the clavicles.
2. Reflect the skin from the neck region, keeping fur away from the dissection field. Pin the head firmly, ventral surface uppermost, on a dry Sylgard-lined dish with four pins.
3. With a small Spencer-Wells artery clamp, reflect the exposed trachea and oesophagus back rostrally, leaving the instrument to maintain the position under its own weight.
4. Under a dissecting microscope, locate the cut ends of the carotid arteries on each side. They will lie superficially and laterally. The vagus nerve and

the thin, sympathetic preganglionic trunk run close alongside, and the ganglia are immediately rostral, but a little deeper.

Either

5. Find the SCG by closely following the course of the carotid artery as it dives deeply to meet the cranium.
   - The ganglion overlies, and is tightly associated with, the carotid artery bifurcation (forming internal and external carotid arteries). It is a plump, white, rather elongate structure, with a very thin preganglionic trunk.
   - The post-ganglionic trunk runs alongside the internal carotid artery, attached to it by a connective tissue sheath, and enters the cranium along the same course as the artery, through the carotid canal.

6. Expose the ganglion more clearly by removing (by tearing out from the rostral end with fine forceps) the longitudinal muscle bands that lie midline but bilaterally, overlying the cervical vertebrae. Resect the ganglion using a pair of No. 5 watchmakers' forceps and a fine pair of spring iridectomy scissors (handling the ganglion only by the nerve trunks).[a]

Or

7. Find the nodose ganglion by locating the vagus nerve running alongside the carotid artery.
   - The vagus nerve is very thick in comparison with the preganglionic nerve of the SCG. Just prior to the carotid artery bifurcation and adjacent to the SCG, the vagus nerve takes a slightly more caudal route, remaining free of the artery and entering the cranium through the jugular foramen.
   - The nodose ganglion is an indistinct swelling on the thick vagal trunk and occurs immediately before the vagus nerve enters the cranium. It is far less evident as a ganglion than the SCG.

8. As for the SCG, expose the ganglion more clearly by resecting the central longitudinal muscle bands which overlie the lower cranium and first cervical vertebrae. Resect the nodose ganglion by cutting with fine spring scissors very close up against the cranial surface. Use forceps to handle only the peripheral vagus nerve trunk.[a]

9. On occasion the ganglion itself is scarcely evident until the thick nerve sheath is removed. If the ganglion is not visible initially, remember which end of the resected nerve was the central end, or mark the peripheral end by tearing the sheath.

---

[a] Dissection of either of these pairs of ganglia should take only 5 min for the two ganglia. During most of this time, the ganglia themselves are protected against drying by the surrounding tissues and the final exposure just prior to resection should last only about 10 sec.

2.4 Culture of ganglion cells

Protocol 2 is used for SCG cells from rats aged 17 days, and works as well for nodose ganglia. For older animals or for newborn animals, adjustments need to be made in the timing and concentrations of enzymes and in the substrates for plating.

Protocol 2. Culture of SCG cells

Materials

- 70% ethanol to sterilize instruments (30 min in 70% ethanol)
- six culture dishes or glass coverslips with a 200 µl blob of laminin in the centre of each, prepare at least 2 h before use
- enzymes: collagenase and trypsin (see Section 2.2.2), each in a 2 ml tube
- two 35 mm culture dishes with 2 ml each of L_{15} medium, which should be kept on ice

Method

1. Following *Protocol 1*, dissect four ganglia. Place them in one of the dishes of L_{15} medium.
2. Using No. 5 watchmakers' forceps (in perfect condition), desheath the ganglia carefully, holding the nerve trunk and peeling back the sheath. Trim off axon trunks, especially thin trailing strands. The desheathing can seem difficult initially, but it is worth gaining the expertise, since it results in a much cleaner final cell culture.
3. Transfer the cleaned ganglia into the second dish of L_{15} medium and cut each one into four roughly equal pieces using fine scissors or a razor blade.
4. Transfer the ganglion pieces into a 2 ml tube, remove the medium, wash once with (Hanks' calcium- and magnesium-free solution (CMF)), then replace with collagenase solution.
5. Leave in the incubator at 37 °C for 15 min. The CMF is not essential for the action of the enzyme at this stage but its addition now allows 15 min for CMF to penetrate the tissue; this extra time improves the final dissociation.
6. Remove the collagenase solution. Gently replace with 2 ml CMF and wash at least twice to ensure removal of divalent cations. Replace with 2 ml trypsin solution. Incubate for 30 min, turning the ganglia over every 10 min.
7. Fire polish three Pasteur pipettes to just visible tip apertures (internal diameter about 0.5 mm). Even with experience it is almost impossible to

judge by eye the internal tip diameter of the perfect pipette. Flame three to varying degrees and judge by their performance.

8. Remove most of the enzyme. Replace with 1 ml culture medium[a] to stop remaining enzyme activity.

9. Replace the culture medium with 2 ml L_{15} medium, and immediately dissociate the cells very gently by passing them through a fire-polished Pasteur pipette. Do not allow air bubbles to pass through the pipette. *This stage is critical to the final condition of the cells.*

   - Remember that gentle handling is very important.
   - Flame the Pasteur pipette to a tip diameter which just allows passage of the ganglion pieces with slight hesitation. They should not become stuck and require extra force to draw them through, nor should they pass so readily as to show no hesitation.
   - Aim for a low cell yield. Remember that the enzyme treatment is too short to achieve complete dissociation.
   - Only two or three passes through the pipette should be needed to achieve the desired slight clouding of the medium (seen only with a light behind the tube).
   - Obvious clouding indicates higher cell yield but may also mean a large-scale if not total loss of the larger ganglion cells, and a sustained change in the condition of the majority of remaining cells. They may grow well, but they may lack their full complement of membrane currents.
   - This dissociation leaves many of the ganglion pieces intact and provides a yield of around 1000 cells per ganglion from a possible 30 000.

10. Prepare a 10 ml tube with 8 ml L_{15} medium. Transfer the dissociated cell suspension, omitting obvious ganglion pieces, into this tube and centrifuge at $500\,g$ for 3 min only.

11. Rinse the laminin-prepared culture dishes with L_{15} very carefully, so as not to wet the whole dish (this might take practice). Otherwise use glass rings to position cells in the centre of the dish.

12. The centrifuged cells form a very loose pellet. Carefully draw off the supernatant and resuspend the cells in growth medium. The volume is governed by the number of dishes prepared.

13. Plate out using 200 µl per dish. Leave in the incubator and allow 3–4 h for the cells to adhere. In a well humidified incubator the 200 µl droplet should be perfectly adequate to maintain the cells for 24 h.[b]

[a] Includes divalent cations and serum, which contains a_2-macroglobulin: a trypsin inhibitor.
[b] The whole procedure from rat to incubator should take a little less than 2 h.

2.5 Critical observation of cells

Immediately following dissociation, the SCG cells should have long, free, tapering dendrites (not contracted nor resembling cudgels) and be well scattered, not clumped. All cells should look healthy and phase bright. Observe the cells again 1 h after plating. They should have resorbed their dendrites and rounded up. Look for phase-dark or lysed cells—there should be almost none: their presence usually indicates either rough handling or low enzyme activity before dissociation. After 3–4 h on laminin, the cells should have adhered (*Figure 2*) and, in the presence of NGF, neurite extension will be beginning. Cells must continue to be phase bright and non-granular. At 24 h the cells will be once again clearly multipolar with well-established neurites of several hundred micrometres; the cells will no longer have the electrophysiologically ideal rounded form (*Figure 2*).

For our purposes the spherical form is the ideal one. To appreciate this, consider what happens as a cell extends dendrites. A perfectly spherical cell of 20 μm diameter would have a volume of 4190 μm^3 and a surface area of 1257 μm^2, with a capacitance of about 12.6 pF. If this cell were to grow four idealized dendrites, each 100 μm long and 1 μm diameter, its surface area will have doubled and its volume increased by 7%. Most of the additional surface area would never be isopotential with a voltage-clamped cell soma due to the high internal resistivity of the dendrites. So how much can we afford to interfere with neuronal development to suit our needs? That depends on what we are trying to measure and what degree of adhesion is needed, because the way in which we secure adhesion can affect the cell's membrane currents. Cell adhesion molecules are definitely not inert sticking devices.

2.6 Control of cell adhesion and neurite outgrowth

Good cell adhesion to a substrate is usually accompanied by promotion of neurite outgrowth. Two main features are important:

- the physicochemical properties of the cell surfaces, such as charge and hydrophobicity
- a variety of receptor–ligand-type interactions involving integral membrane adhesion proteins

Where ganglion cells are concerned, we are dealing mainly with a superfamily of these proteins: the integrins. These interact with complex glycoproteins of the extracellular matrix and basement membranes, e.g. laminin (which has a binding K_d of 10^{-9} M), fibronectin, and collagen. Adhesion proteins can have profound effects, not only on the cell's capacity to adhere, but also on gene expression, enzyme induction, migratory capability, neurite outgrowth, and even cell survival. The significance of these adhesion proteins and their variety differs between cell types and also with

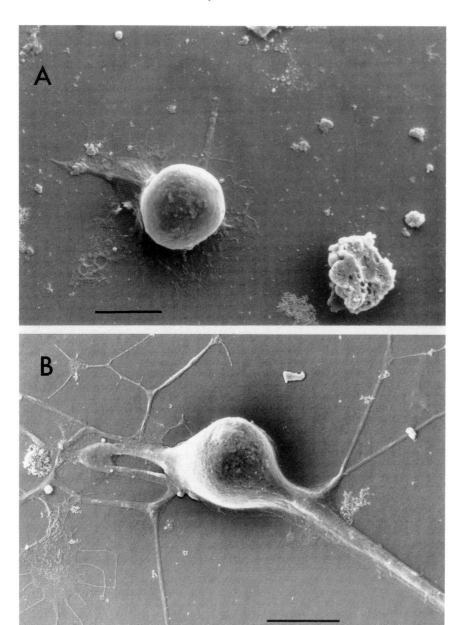

Figure 2. Rat SCG neurones cultured on laminin and fixed at 3 h (A) and 16 h (B) after plating. Processing for the scanning electron microscope has resulted in considerable (approximately 50%) shrinkage of the cell somas. Bars represent 10 μm. Most of the living cells are normally around 30 μm. For voltage-clamp purposes the spherical form is ideal.

the stage of cell maturity. Thus, from the moment of isolation the cell is subject to change, and the method we choose to substitute for these disrupted receptor–ligand interactions can induce further rapid change in the state of the cells. It is worth testing various conditions for cell adhesion to optimize the particular phenomena we wish to study.

Glass or plastic surfaces will need precoating with either synthetic polycationic macromolecules, such as polylysine, or the adhesive glycoproteins mentioned above: laminin or collagen. Serum is usually required in the growth medium for the first few hours after plating to encourage adhesion.

To limit the rate of neurite growth, suboptimal concentrations of substrate could be tried or a change of substrate, e.g. SCG cells from 17-day-old rats extend fewer neurites on a collagen substrate than on laminin, but they also take longer to adhere, so there is a greater degree of re-aggregation. NGF in the growth medium could be reduced from 10 to 3 µg/ml or omitted altogether.

3. Recording from ganglion cells

3.1 Choice of voltage-clamp amplifier

Considering the size of the currents encountered in the ganglion cells, the correct approach is to use a two-electrode voltage-clamp. However, for general purposes, understandably, most researchers prefer to make do with a single electrode. The choice of amplifier rests on the interests of the researcher and there are two main types: (a) patch amplifier and (b) switching amplifier.

The switching amplifier uses a technique described by Wilson and Goldner (8) in which the function of the single electrode is switched rapidly, at 3–20 kHz, between voltage-recording and current-passing modes (with options for continuous single-electrode voltage-clamping when required, and also for two-electrode voltage-clamp). In contrast to the patch amplifier, in which the clamp feed-back amplifier is an I/V converter, this amplifier is a voltage-follower. For work with a single electrode, the switching amplifier has two main advantages over the patch amplifier:

- Where large currents are involved, series resistance errors are avoided because no current passes through the electrode at the moment of voltage measurement.
- The same system can be used for either patch pipettes or for fine, high-resistance impalement electrodes. This, at times, is important since the cellular perfusion achieved with a patch pipette may disguise or even occlude the cellular response under study, and the option to compare results with the two methods is a very valuable one.

This type of amplifier is also a good current-clamp amplifier. Its disadvantage is that the current record is noisy, and clamp speed limited by the switching frequency. Small currents with slow kinetics may be filtered satisfactorily, but rapid signal components will be lost.

The patch amplifier uses the electrode for simultaneous current passing and voltage recording and can be used for both macroscopic (whole-cell) voltage-clamp currents and for higher amplification single channel studies. The low noise recording, in comparison to the switching amplifier, makes it a good choice for the study of currents below a nanoamp and enables a wider bandwidth for the study of very fast events, especially if an integrating headstage is used (Axon Instruments). However, the technique must be handled carefully to avoid series resistance errors.

3.2 How well can these cells be clamped?

Voltage control in these cells can be excellent so long as experiments are designed within the limitations of the recording systems employed. The question of isopotentiality will have been largely dealt with if we choose to work on well-rounded cells, but more complex problems arise when we consider the density of the expected currents and the total membrane capacitance (C_m).

Ganglion cell size (10–60 μm) is only a rough guide to the membrane capacitance and surface area. For perfect spheres the corresponding range of membrane capacitances would be 3–113 pF (1 μF/cm^2) (*Figure 3A* and *B*), but due to non-ideal cell shape, estimates of surface area should be made from capacitance measurement.

The sizes of active ionic currents vary enormously, but in a standard Krebs' solution the potassium currents tend to be the largest. In our hands, when depolarizing the membrane from −80 mV (i.e. without recruiting the whole population of channels, some being inactivated), SCG delayed rectifier or the DRG A-type currents confer a maximal conductance (G) of around 2 nS/pF. Calculating for the ideal spherical membrane (since $I = G \times V$), the activation of a single potassium channel species, with 100 mV driving force for potassium, can evoke outward current magnitudes of 2.5 nA in a 20 μm (13 pF) cell, 10 nA in a 40 μm (50 pF) cell and 23 nA in a 60 μm (113 pF) cell (*Figure 3*). If the whole potassium channel population were recruited or if calcium-activated currents were evoked, then the total currents could be considerably larger. The size of these currents means that, using a patch amplifier, uncompensated series resistance (R_s) errors can seriously interfere with voltage-clamp control (see reference 2). The current and voltage traces, however, can look just fine! The problem does not stop here: for effective whole-cell patch technique, R_m should be many times, perhaps two orders of magnitude, higher than R_s, noting that in practice R_s may be 7–10 MΩ, even when starting with a 2 MΩ electrode, Yet as these potassium currents activate,

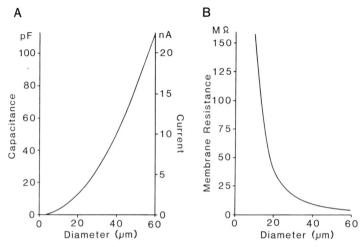

Figure 3. Idealized examples of the relationship between the diameter of a perfectly spherical membrane and some steady-state membrane characteristics that would influence voltage-clamp effectiveness: capacitance, current magnitude, and input resistance. Ganglion cells have a high current density of 2 nS/pF for their delayed rectifier or A-type currents. The idealized example presented above assumes a spherical membrane of infinite resistance apart from an activated, steady-state potassium conductance of 2 nS/pF, with a driving force of 100 mV. The capacitance is assumed to be 1 μF/cm². The membrane time-constant would be 500 μsec. in spheres of all sizes. (A) Relationship of the sphere diameter to its capacitance (left abscissa) and predicted outward potassium current (right abscissa). In a real cell, capacitance is often somewhat greater than might be predicted from its diameter, and gives a more accurate estimate of cell surface area. The relationship of surface area to current magnitude is, of course, linear. (B) Relationship of the sphere diameter to R_m, all cell sizes having the same current density. In this example, the presence of a measuring/clamping electrode has not been considered; however, note that for a voltage-clamped cell with these current magnitudes, R_m could easily reach values similar to that of R_s, even in a 20 μm cell.

the cell's input resistance is predicted to fall from several hundred megaohms to only 40 MΩ in the case of the 20 μm cell, and to merely 10 and 4.4 MΩ in the 40 and 60 μm cells, respectively (*Figure 3*). The result is that the value of R_m would approach that of R_s. This can generate a serious unseen error due to the voltage drop across the pipette, and steady-state clamp errors can be quite large.

In addition, since the time constant, $\tau = RC$, a significant decrease in the membrane charging time constant occurs. Use of the switching amplifier is ideal only if the settling time constant for the electrode voltage is at least 100 times faster than that of the membrane. In the situation posed in *Figure 3*, the membrane time constant falls to only 500 μsec in all cells, irrespective of size, so that in order to maintain a stable membrane voltage, switching frequency would need to be at least 10 kHz. A brief membrane time constant also means that clamp settling time is prolonged. The membrane time constant should preferably be restrained to values greater than 1.5 msec. In other words, full control of the larger potassium currents in these cells requires a two-electrode approach. This, of course, does not preclude the acquisition of a large range of very valuable experimental data which can be obtained using a single-electrode method, and a degree of common sense. See also Section 3.3.2 for other implications of this change in membrane time constant with regard to leak subtraction and tail currents.

To propose some main guidelines:

(a) Remember that 'bigger' is not necessarily 'better'. If the currents are expected to be large, choose small cells whenever possible. Alternatively, use only a portion of the cell membrane by isolating a large multichannel patch. If necessary, modify the ionic solutions to reduce ionic driving force, or limit the voltage range of the study to keep the currents small.

(b) Minimize the series resistance. Use low-resistance recording pipettes (~ 2 MΩ) and watch for signs of increase in R_s during experiments, taking steps to reduce it whenever necessary.

(c) For large currents, avoid the series resistance problem. Either use a switched single electrode voltage-clamp (but remember that here too, the clamp will become ineffective if, during large currents, the membrane time constant is allowed to fall to ranges close to that of the electrode itself), or by using a two-electrode voltage-clamp method.

(d) Keep electrode capacitance down by using low bath fluid levels, by coating electrodes with Sylgard, and by using thick-walled glass where appropriate.

3.3 Leak—and how to deal with it

3.3.1 What is leak?

'Leak' is a general term which includes several phenomena simultaneously, some well-defined, others very ill-defined. Firstly, in the recorded cell 'leak' refers to a low-resistance pathway at the sealed junction of glass and cell membrane, most particularly in penetrated cells. In patched cells this junctional resistance ought to be at least a gigaohm. This leak is generally assumed to show no ionic specificity and is linear with respect to voltage, with a reversal potential at 0 mV. However, the cell membrane may also have other unidentified 'leak' resistances due to persistent ion channel activity, which may or may not be specific to an ion species. If, for example, internal free calcium is raised above 100 nM, then calcium-activated chloride or potassium conductances (or both) may be tonically active. This channel activity will contribute a 'leak' component, which may also be linear with respect to voltage, but with a reversal potential which accords with that of the charge-bearing ionic species. There are many such underlying conductances in different cells, in some cases probably active in the intact cell, in others active only under specific recording conditions; we may need to identify them if they interfere with the phenomena we wish to study.

Avoidance of inadvertently calcium-loading the cell is most important. Loading often occurs on initial penetration, or by momentarily disturbing the patch seal on entering whole-cell patch mode, or during the experimental voltage-clamp protocols. Calcium readily enters through two routes: (a) through the glass–membrane junction, and (b) through voltage-gated calcium channels, and can trigger further release from intracellular stores. Such entry readily overloads the buffer capacity of the cell and that provided by the pipette medium, unless high concentrations of calcium chelators are used. Yet the use of high EGTA or BAPTA concentrations (e.g. 10 mM) can be counter-productive, if one is studying phenomena which are dependent on calcium signalling; e.g. 0.1 mM might be more appropriate. To avoid problems with intracellular calcium the minimal rules are:

- aim for a high pipette–membrane seal resistance (> 1 GΩ)
- keep the cell hyperpolarized, to keep calcium channels closed, e.g. enter whole-cell patch configuration in voltage-clamp mode with the voltage pre-set to -60 mV
- take care when designing experimental protocols to avoid progressive calcium-loading

3.3.2 Leak subtraction

Computer-assisted analysis of voltage step records can make leak subtraction easy and quick. Using a standardized low-voltage step from the linear portion of the *I/V* curve in which no obvious currents are active, the resultant current and capacitative transient can be scaled (by scaling the leak voltage to match that of the test pulse) and subtracted, assuming linearity (note the assumption) of the leak and capacity. If some effort has been made to limit the number of active currents in the cell (e.g. by isolating calcium or potassium currents), then this leak subtraction can work very well indeed. However, if we are interested in the kinetic behaviour of ion channels, then problems can arise even when restricting ionic movement to a single ion species. The problem arises not with the leak, but with the subtraction of the capacity transient. As described by Finkel (2), the membrane-charging time constant is $\tau = R_p C_m$, where R_p is the summed values of the access resistance and membrane resistance. A voltage step from a region of low conductance to a voltage that evokes substantial membrane current will often show an asymmetric capacity transient in the current trace at the onset and offset of the voltage step because of the difference in membrane resistance. Thus, an assumption of linearity of the capacity transient for purposes of subtraction is erroneous, and interpretation of fast deactivating tail currents should be made with due caution.

4. Solution design and adequacy of exchange between pipette and cell interior

4.1 General points

The choice of solutions, not surprisingly, depends on which currents are to be studied. Whole-cell recording allows control of both the extracellular and intracellular solutions, so some thought must be taken on what is required. The extracellular solution is usually fairly straightforward; normally it takes the form of a conventional Ringer's or Krebs-like solution, with perhaps tetrodotoxin (TTX) or other pharmacological agents added to block particular channels when required. It is of interest that, in sympathetic neurones, it is necessary to use relatively high concentrations of TTX (0.5–1 µM) to get complete block of the sodium current. Lower concentrations often leave a residual transient current which is dangerously similar to a transient calcium current.

The intracellular solution is much more problematic. Apart from choosing the major ions, it is also necessary to control the intracellular pH and calcium concentration. Furthermore, while whole-cell recording causes dialysis of the cell which allows the intracellular solution to be controlled, it also means that essential intracellular constituents may dialyse out of the cell into the pipette, changing the physiological response of the cell. This is often manifested in the gradual (sometimes not so gradual) rundown of the currents of interest.

4.1.1 Composition of the intracellular solution

The simplest intracellular solution consists of KCl (normally about 150 mM) with some MgCl (perhaps 2 mM), plus a calcium and a pH buffer (perhaps 10 mM EGTA and 10 mM Hepes). Often K^+ is replaced by Cs^+ to block potassium currents when other currents are studied.

Another variation often used is largely to replace the anion, Cl^-, with aspartate. This tends to give more stable recordings, but exacerbates a problem that is usually not large when swapping cations. This is that the different mobilities of Cl^- and aspartate in solution gives rise to a significant junction potential between the two solutions.

4.1.2 Junction potentials

During whole-cell recording the solutions are separate; however, before forming a seal they are in contact and a junction potential exists. This is normally balanced out as part of the offset current. On forming a seal, the solutions are isolated from each other and the holding voltage is then erroneously offset by the amount of the junction potential. One can easily measure the junction potential between solutions with a ceramic junction reference electrode, filled with a saturating concentration of KCl, and correct accordingly. For example, the junction potential between an internal solution with 150 mM KCl and one with 150 mM potassium aspartate is about 10 mV. Whether this means that when a cell is apparently held at −60 mV, it is actually at −50 mV or at −70 mV, makes for an exciting and animated laboratory party game!

4.1.3 Osmolarity

It is also important to measure the osmolarity of solutions when large changes in solution composition are made. This can be done quickly with a vapour pressure osmometer, which is easy to use but more difficult to calibrate. The normal osmolarity of the external solution is between 300 and 330 mOsm. Often, whole-cell recordings last longer if the internal solution is kept slightly hypo-osmotic (by about 10%) compared with the external solution.

4.1.4 pH and intracellular calcium concentration

The buffering of intracellular pH and calcium is critical. For calcium, it may be desired to buffer strongly, or to hold calcium at a certain concentration and yet still allow it to rise in response to certain hormones. EGTA or BAPTA at 10 or 20 mM will do the former, while 0.1 mM EGTA or BAPTA alone will allow calcium to rise. For pH, 10 mM Hepes is usually sufficient. Titration is usually carried out with KOH, NaOH or CsOH as appropriate.

It is also essential to filter the internal solution before trying to form a seal; it is often necessary to filter the external solution too, if sealing seems unusually difficult.

4.2 Time course of cell dialysis

It is important to remember that dialysis of the cell is not instantaneous; the rate of equilibration of a given molecule between the electrode and the inside of the cell will depend on the molecular weight of the molecule, the series (access) resistance between electrode and cell, and the volume of the cell (proportional to its capacitance). For example, Beech et al. (9) found that the mean capacitance of freshly isolated sympathetic neurones was 36 pF—quite large but not excessively so. In these experiments, the series resistance was fairly low (3.5 MΩ) as wide-tipped/low-resistance pipettes were used. Fura-2 uptake into these cells was found to occur with a time constant of 399 sec. The uptake of any other molecule into the same population of cells can be calculated from the equation:

$$\tau_x = \tau_f (M_x/M_f)^{1/3} \quad (1)$$

where τ_x is the time constant of the unknown molecule, M_x its molecular weight, τ_f the time constant for fura-2 and M_f its molecular weight.

Thus, BAPTA in these cells would be taken up with a time constant of about 362 sec. In other words, if the pipette contains 20 mM BAPTA, the cell will reach a concentration of 10 mM after about 5 min of dialysis. This is a considerable length of time to remain immobile, staring at the oscilloscope screen!

In the absence of fura-2 measurements, it is possible to estimate the time constants for various molecules by extrapolation from published data in other cell types (providing one knows the mean capacitance and series resistance for the cells being studied), using the following equation:

$$\tau_u = (R_u C_u^{3/2}/R_k C_k^{3/2}) \times \tau_k \quad (2)$$

where R_k and C_k are the series resistance and capacitance in the cell with known rates of accumulation, τ_k is the time constant for the molecule in those cells, τ_u is the unknown time constant for the same molecule into the cell (with series resistance and capacitance R_u and C_u) where the rate of accumulation is unknown.

A comprehensive study of the uptake of different molecules in bovine chromaffin cells was carried out by Pusch and Neher (10) and provides useful numbers to use as standards. Thus, one can use Equation 1 to estimate the uptake of any molecule into chromaffin cells and then Equation 2 to calculate the equivalent uptake into any cell. It must be emphasized, however, that this type of calculation gives only rough 'ball-park' figures and that it is dangerous to extrapolate calculated rates to cells with vastly different sizes. Chromaffin cells are fairly small with a capacitance of around 6 pF. Some examples of predicted uptake rates for different-sized molecules into a theoretical sympathetic neurone are shown in *Figure 4*.

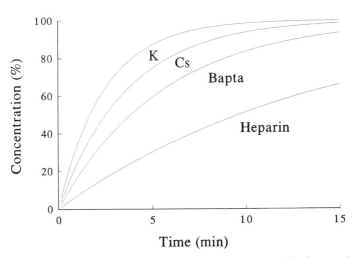

Figure 4. Uptake of four molecules of different molecular weight into a theoretical sympathetic ganglion cell with whole-cell capacitance of 30 pF. The access resistance was set to 5 MΩ. Values were calculated using Equations 1 and 2, together with the measured uptake rate for Fura-2 into a population of sympathetic ganglion cells obtained by Beech and colleagues (9).

4.3 Rundown of currents during whole-cell and single channel recording

As mentioned above, whole-cell recording allows molecules to dialyse out of, as well as into, the cells, which can lead to a rundown of labile currents and a loss of certain modulatory pathways which activate diffusible messengers.

4.3.1 Additions to the pipette solution

A number of things can be added to the intracellular solution to help prevent this rundown of activity. Mg ATP (2–3 mM) is essential, both to maintain labile currents and to play a permissive role in allowing modulation of currents by neurotransmitters. Many laboratories also add GTP (0.1 mM), although this seems less essential. Note that each of these nucleotides have to be added freshly each day to the intracellular solution. Others (e.g. reference 11) add an ATP-regenerating system consisting of dibutyryl cyclic amp (cAMP) (5 mM), the catalytic subunit of cAMP-dependent protein kinase (about 20 µg/ml) and the protease inhibitor leupeptin (80–100 µM) which in some systems leads to calcium currents that are stable for several hours. This approach is probably not necessary for every occasion, but is worth trying if rundown problems persist. In previous work (9), one of us routinely added the protease inhibitor leupeptin to the intracellular solution because it had become a custom to do so. The author remains unconvinced that this made a great deal of practical difference to the maintenance of currents and their modulation in sympathetic ganglion cells; however, psychologically it worked remarkably well!

Many of these treatments work well in whole-cell recording to give stable currents and responses. They work with much less success, however, in isolated patches, particularly in outside-out patches where, for example, it is still extremely difficult to record individual calcium channels for any length of time.

4.3.2 The perforated-patch recording technique

In some cases, to prevent rundown in whole-cell recording, it may be necessary to use the technique of perforated-patch recording developed by Horn and Marty (12; see also Chapter 2). In this method, the antibiotic nystatin is added to the pipette solution and a seal is formed as normal. With luck, the nystatin in the pipette solution will enter the cell membrane and form channels. This allows electrical access to the cell without dialysing its contents. Protocols which work regularly (particularly in terms of nystatin concentration) seem to vary quite a bit from cell to cell. *Protocol 3* has given some success with rat sympathetic neurones.

Protocol 3. Preparation of nystatin

1. Make up a nystatin stock solution (5 mg/100 µl dimethylsulphoxide (DMSO)). Vortex, then sonicate for ⩾ 15 min.
2. Make up a pluronic stock solution (25 mg/ml DMSO). Heat for ⩾ 10 min at 37 °C.[a]
3. For 1 ml of internal solution, add 10 µl nystatin (500 µg/ml) and 20 µl pluronic (500 µg/ml).
4. Vortex, sonicate the internal solution for ⩾ 5 min and then filter (there will be yellow debris on the filter).
5. Use this internal solution within 1–2 h.

[a] Note that pluronic seems to act as a dispersant and helps to get nystatin into the solution.

This method of recording has its problems. The success rate is generally lower than with conventional whole-cell recording and the increase in access resistance can lead to quite severe voltage-clamp problems when dealing with large and fast currents, e.g. sodium or calcium currents. However, when it works, it works extremely well. Recordings can be much longer than with conventional whole-cell recording and currents, and agonist responses are often much more stable and reproducible. Recently, Levitan and Kramer (13) have tried to extend this technique to outside-out patches to combat rundown there. They seem to have had good success with patches from cell lines (GH_3 cells). However, it remains to be seen if the technique can be extended to patches from mammalian neurones.

4.4 Other considerations

It is clear that there is a lot to think about when choosing solutions to optimize whole-cell and single channel recordings.

It is also worth noting that there are sporadic reports in the literature of the choice of filter-type or the choice of electrode glass altering the properties of the channels of interest, often because divalent cations can leach into the intracellular solution. For example, there is a recent suggestion that there may be a leakage of Ba ions from a particular type of soft glass into the intracellular solution in the patch pipette (14) at concentrations which may begin to block certain potassium channels.

Having said that, most techniques have been fairly comprehensively tried and tested nowadays. The division between being a good, careful experimenter and being what Americans call 'anal' is still fairly clear!

5. Channels and receptors

It is beyond the scope of this chapter to review the literature concerning the ion channels and receptors of peripheral ganglion cells. Instead, we have chosen to show some examples of the currents that can be recorded in mammalian sympathetic ganglion cells (see *Figures 5* and *6*) and concentrate on two channel types in particular, calcium channels and potassium channels. We discuss in some detail how best to study them using whole-cell and single channel recording.

5.1 Calcium currents

5.1.1 Stable recordings of calcium current

To record stable calcium currents in whole-cell mode, it is usually necessary to add Mg ATP to the pipette solution (see above). Additionally, calcium entry through voltage-dependent calcium channels leads to calcium-dependent inactivation of the current, which causes it to appear to run down with time. Steps need to be taken to avoid this. There are a number of alternatives. One can use Ba^{2+} as the charge carrier rather than Ca^{2+}, load the cells with a calcium buffer, or elicit the current with short-lasting steps or with a large delay between pulses to minimize calcium entry. It seems to be the case that certain modulatory effects on calcium currents (e.g. by activation of muscarinic receptors) may be altered when the intracellular calcium concentration is highly buffered. An alternative protocol, which leads to little rundown of the calcium current in rat sympathetic neurones, is to use 5 mM Ca^{2+} outside as the charge carrier, leave the intracellular calcium concentration free to rise with only 0.1 mM BAPTA, but use short 10–12 msec depolarizing steps (to 0 mV from a holding voltage of −80 mV) once every 4 or 5 sec. This gives a whole-cell current that runs down very little over the course of a short experiment (10–15 min).

5.1.2 Isolation of calcium currents

The calcium current is isolated from contaminating potassium currents by loading the cells with caesium. Leak subtraction isolates it further from any residual potassium currents or leak currents. A good way to achieve this is to perfuse an external solution containing 100 μM Cd^{2+}. This will block the entire calcium current selectively, quickly and reversibly; the resulting response to depolarizing steps can be subtracted from the control current.

The alternative method of subtracting the current produced by a scaled hyperpolarizing pulse, described earlier, will only account for any linear leakage component.

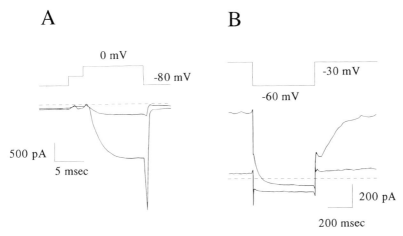

Figure 5. Examples of whole-cell currents from sympathetic ganglion cells. (A) Whole-cell calcium current and its modulation by muscarinic receptor activation with 10 μM oxotremorine-M. (B) Whole-cell M current, also modulated by 10 μM oxotremorine-M. In each case the dashed line represents the zero current level. (From unpublished records of L. Bernheim and A. Mathie.)

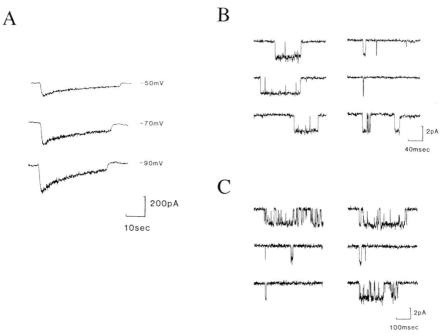

Figure 6. Examples of ligand-gated ion channel currents in rat sympathetic ganglion cells. (A) Whole-cell nicotinic currents evoked by 20 μM ACh. (B) Sample records of single nicotinic channels in an outside-out patch activated by 20 μM ACh. (C) Sample records of single-channel currents evoked by activation of $GABA_A$ receptors by 20 μM GABA in an outside-out patch.

5.1.3 Separation of calcium current into N, L, T, and P

Dissecting the whole-cell calcium current into N, L, T, or P is tricky. Different groups prefer different methods (15). T currents are generally the easiest to separate, as they are activated by much smaller depolarizing voltages than the others; hence, they are also known as low-threshold calcium currents. It is generally agreed that there are few or no T-type calcium channels in rat sympathetic ganglion cells. P-type channels seem, at present, to be restricted to the central nervous system. Thus, in sympathetic neurones, we are left with the problem of distinguishing N from L. Often, this is done pharmacologically because N-type channels are said to be sensitive to omega-conotoxin and L-type channels are sensitive to dihydropyridine (DHP) agonists and antagonists. The whole-cell calcium current in rat sympathetic neurones is blocked by about 80% by omega-conotoxin, suggesting it is mostly carried through N-type channels. However, there are certainly L-type channels present. The best way to observe them is to look at the prolongation of tail currents by a DHP agonist (such as (+)202–791).

There is a suggestion that, in some cell types, the effects of omega-conotoxin and DHP are not clear cut and there is some overlap between their actions. However, at present, for rat sympathetic neurones, it seems the best method of separation. The alternative method of classifying the current by the degree of inactivation of the current (with N-type channels inactivating and L-type channels not) simply will not work, since N-type channels have been shown to uncouple from inactivation to give periods of sustained activity in rat sympathetic neurones (16). Others swear by single-channel properties for separation for N and L. This works for some cell types (such as DRG cells), where the conductance of L- and N-type channels are clearly different. However, this is not an appropriate method in rat sympathetic neurones, where the conductances of N- and L-type calcium channels are quite similar.

5.1.4 Single channel calcium currents

Because of rundown problems, single channel calcium currents can only really be studied successfully in cell-attached patches. The best method is to fill the pipette solution with 110 mM $BaCl_2$ and 10 mM Hepes. The individual channels have a conductance that is too small to work with if calcium is the charge carrier. The bath solution is more complex. A typical solution might contain high potassium (perhaps 155 mM potassium aspartate) to 'zero' the membrane potential of the cell, so that one can voltage-clamp the patch correctly. It must also then contain a calcium buffer (5 mM EGTA say) to stop calcium entry into the cell and cell death.

The voltage protocol has to be altered also. Usually a depolarizing pulse with a minimum duration of about 100 msec is required to see a sufficient number of individual openings in the course of an experiment. The major drawback with cell-attached patches is that it is difficult to change the external solution around the channels once the seal is formed. It is possible, but quite tricky, to perfuse the patch pipette. Luckily, DHPs can be applied to the bath, as they can diffuse across the cell membrane. However, if for example omega-conotoxin is to be studied, it would have to be added to the pipette solution before the experiment was begun and control activity obtained from a separate patch. This is often less than satisfactory and explains the continued attempts of certain groups to record maintained calcium currents in outside-out patches.

5.2 Potassium currents

Potassium currents are the most diverse type of currents identified so far in mammalian peripheral ganglion cells, as in most neurones. We will consider five potassium channel types here: the delayed rectifier I_{DR}, the A current I_A, two calcium activated currents $I_{K(Ca)}$ and $I_{K(AHP)}$, and the M current I_M. These are all found in rat sympathetic neurones, but there may certainly be more.

5.2.1 I_A, I_{DR}, and $I_{K(Ca)}$

From a resting potential of −60 to −70 mV, depolarizing steps elicit potassium currents made up of at least three components. These components can be looked at in relative isolation using *Protocol 4*. The protocol is adapted from experiments originally described by Belluzzi, Sacchi, and Wanke (see 17 and references therein) using neurones in intact ganglia recorded with microelectrodes. It is of interest that these experiments, though less versatile than those with patch-clamp recording methods, give currents that are much larger. I_{DR} and I_A can be 20–40 nA and the calcium currents from these neurones are 10–15 nA at their peak, compared with calcium currents of 1–2 nA in patch-clamp recordings of isolated neurones. Much of this discrepancy is undoubtably due to the different techniques selecting different populations of cells; microelectrode experiments require large-diameter cells, while it is better to pick small-diameter cells for patch-clamp experiments—see Section 3.2.

Protocol 4. Simple separation of I_{DR}, I_A, and $I_{K(Ca)}$

1. Hold the cell at −90 or −100 mV. (Alternatively, pre-pulse to −100 mV for at least 1 sec from a more depolarized holding voltage.)
2. Step to various potentials (range −60 to +30 mV) for about 50 msec once every 10 sec.
3. Hold the cell at −50 mV (or pre-pulse as above) and repeat the same steps.
4. Subtract currents in (3) from those in (2). This gives I_A.

Protocol 4. *Continued*

5. Hold cell at −50 mV and step (from −30 mV to +30 mV) for 200 msec once every 15 sec.
6. Repeat (5) in the presence of 0.5 mM Cd (to stop Ca entry through Ca channels). This gives I_{DR}.
7. Subtract currents in (6) from those in (5). This gives $I_{K(Ca)}$.

The A current decays rapidly with a time constant of around 10 msec (although this varies with temperature and voltage step), so longer steps are not required. I_{DR} and $I_{K(Ca)}$ inactivate only slowly and take longer to activate. Thus, 200 msec steps are required to get close to full activation of the current. Note that $I_{K(Ca)}$ is contaminated by I_{Ca}, for which a correction may need to be made. Furthermore, it should be borne in mind that such a protocol, with prolonged depolarizations, may cause significant calcium loading of a cell, with the consequence of possible alterations in the calcium-dependent current as a result.

Pharmacological separation of these potassium currents using, for example, tetraethylammonium chloride (TEA) and 4-aminopyridine (4-AP) is less advisable, because there seems to be some overlap between the sensitivity of different channels to different agents.

5.2.2 $I_{K(AHP)}$

A second calcium-activated potassium current, $I_{K(AHP)}$, requires a different style of protocol to be seen in isolation. This current deactivates slowly on repolarizing the cell, following its activation by depolarization. This feature can be exploited by depolarizing the cell to 0 mV for 50 to 100 msec, then looking at the tail current at −50 mV for 100 msec or so following this. In contrast to $I_{K(AHP)}$, $I_{K(Ca)}$ completely inactivates within a few msec at this potential. Pharmacologically, $I_{K(AHP)}$, unlike $I_{K(Ca)}$, is sensitive to apamin. At the single channel level, $I_{K(Ca)}$ may be due to the opening of large (100–300 pS) BK channels, while $I_{K(AHP)}$ corresponds to the opening of much smaller conductance SK channels. Separation of each of the different potassium channels at the single channel level is, however, much more difficult than in the whole cell, as it is extremely difficult to record a single potassium channel type in complete isolation. This is reflected in the literature, where only BK channels of sympathetic ganglion cells have been studied at the single channel level in any depth so far.

5.2.3 M current (I_M)

Before leaving potassium currents, it is necessary to consider the M current. Although small (usually less than 1 nA), it is important because it switches on as the cell is depolarized, thereby acting as a negative feedback control of excitability (18). It is, however, inhibited by activation of muscarinic receptors; thus, it is switched off by released acetylcholine (ACh) during periods of sustained synaptic transmission through nicotinic receptors. This current can be studied in relative isolation because it does not inactivate. Normally, the M current is activated by holding a cell at −30 mV (see *Figure 5*). The cell can then be stepped to −60 mV for 0.5 to 1 sec every 5 to 10 sec where the M current deactivates (switches off) slowly with a time constant of around 150 msec. The M current can be quantified precisely, by taking the difference between the instantaneous current at −60 mV following the step (when the channels are still activated) and the steady-state current at −60 mV, 500 msec later, when all the channels have closed. Like calcium currents, this seems to be a labile current so that similar precautions have to be taken to avoid run-down of the M current with time.

5.3 Neurotransmitter receptors and ion channels

Finally, it is worth noting that these cells possess a number of different receptors, which are either ligand-gated ion channels or control cell excitability by modulating ion channels.

5.3.1 Ligand-gated ion channels

Every cell possesses nicotinic ACh receptors and γ-aminobutyric acid ($GABA_A$) receptors. About 30% of cells have glycine receptors and cells with large diameter (87% of cells with capacitance more than 30 pF) possess $5\text{-}HT_3$ receptors.

5.3.2 Modulation of ion channels

Additionally, cells possess muscarinic receptors which modulate I_{Ca} and I_M; noradrenaline receptors which modulate I_{Ca} and $I_{K(AHP)}$; and somatostatin, adenosine, and dopamine receptors which also modulate I_{Ca}.

5.4 Summary

It seems certain that this list of receptors and ion channels is not complete and that much work remains for those interested in studying the excitability of sympathetic and other peripheral ganglion cells using patch-clamp recording methods.

Acknowledgements

We would like to thank Prof. David Brown for his encouraging comments and criticism of the manuscript. For the images of cultured cells in Figure 2, we thank Yvonne Vallis (Department of Pharmacology, UCL) for the cell culture, and David McCarthy (E. M. Unit, School of Pharmacy, University of London) for the electron microscopy. The images are from a Philips XL20 scanning electron microscope.

References

1. Lindsay, R. M., Evison, C., and Winter, J. (1991). In *Cellular Neurobiology: A Practical Approach* (ed. J. Chad and H. Wheal) pp. 1–17. IRL Press, Oxford.
2. Finkel, A. (1991). In *Molecular Neurobiology: A Practical Approach* (ed. J. Chad and H. Wheal) pp. 3–25. IRL Press, Oxford.
3. Dempster, J. (1989). In *Microcomputers in Physiology: A Practical Approach* (ed. P. J. Fraser) pp. 51–94. IRL Press, Oxford.
4. Gurney, A. M. (1990). In *Receptor–Effector Coupling: A Practical Approach* (ed. E. C. Hulme) pp. 155–180. IRL Press, Oxford.
5. Kostyuk, P. G. (1991). In *Cellular Neurobiology: A Practical Approach* (ed. J. Chad and H. Wheal) pp. 121–135. IRL Press, Oxford.
6. Lancaster, B. (1991). In *Cellular Neurobiology: A Practical Approach* (ed. J. Chad and H. Wheal) pp. 97–120. IRL Press, Oxford.
7. Borg-Graham, L. J. (1991). In *Cellular Neurobiology: A Practical Approach* (ed. J. Chad and H. Wheal) pp. 247–275. IRL Press, Oxford.
8. Wilson, W. A. and Goldner, M. M. (1975). *J. Neurobiol.*, **6**, 411.
9. Beech, D. J., Bernheim, L., Mathie, A., and Hille, B. (1991). *Proc. Natl. Acad. Sci. (USA)*, **88**, 652.
10. Pusch, M. and Neher, E. (1988). *Pflügers Archiv.*, **411**, 204.
11. Chad, J. and Eckert, R. (1986). *J. Physiol.*, **378**, 31.
12. Horn, R. and Marty, A. (1989). *J. Gen. Physiol.*, **92**, 145.
13. Levitan, E. S. and Kramer, R. H. (1990). *Nature*, **348**, 545.
14. Copello, J., Simon, B., Segal, Y., Wehner, F., Ramanujam, V. M. S., Alcock, N., and Reuss, L. (1991). *Biophys. J.*, **60**, 931.
15. Tsien, R. W. and Tsien, R. Y. (1990). *Annu. Rev. Cell Biol.*, **6**, 715.
16. Plummer, M. R. and Hess, P. (1991). *Nature*, **351**, 657.
17. Belluzzi, O. and Sacchi, O. (1990). *J. Physiol.*, **422**, 561.
18. Adams, P. R. and Brown, D. A. (1982). *J. Physiol.*, **332**, 263.

2

Transport mechanisms in ocular epithelia

T. J. C. JACOB, J. W. STELLING, AMANDA GOOCH, and
JIN JUN ZHANG

1. Introduction

The epithelia of the lens and ciliary body of the eye represent very different structures both from structural and functional viewpoints. The reasons for studying them are, however, related. Two diseases of the eye, cataract and glaucoma, result from the malfunctioning of these respective epithelia.

All epithelia are involved in transport and, although there are some broad categories, each individual epithelium tends to possess specific faculties that are directly related to its particular function. In the case of the lens, this function is to maintain the ionic and solute balance in the tissue to prevent swelling and opacification, and in the case of the ciliary epithelium, it is to secrete aqueous humour. Although these two epithelia have distinct functions, they nevertheless have a number of channels, exchange mechanisms, and pumps in common, and this chapter deals with the techniques used in their identification.

Previous experimental approaches to both the lens and the ciliary body involved techniques that, although they provided some of the basic physiological information on tissue function, had drawbacks limiting their usefulness. Kinsey and Reddy (1), using a type of whole-lens double (Ussing) chamber, were able to show that the lens acts as a 'pump–leak' system for potassium and amino acids. These substances are actively accumulated by the anterior epithelium, diffuse posteriorly, and leave the lens across the posterior capsule. Sodium moves in the opposite direction and is pumped out at the anterior epithelium (see Davson, (2) for a review). Intracellular recording in the lens has determined that the lens is a coupled syncytium with a uniform potential in the intracellular compartment (3, 4). Ions are therefore free to move from cell to cell following gradients set up by the activity of the Na^+/K^+-ATPase pumps in the anterior epithelium. If these pumps are inhibited with ouabain the lens swells and opacifies (5, 6), but how the epithelium achieves solvent transport has never been determined. Techniques that address this question are described in Section 2.

Double-chamber studies of transport in the ciliary body have shown that the transepithelial potential is negative, i.e. the inside surface is negative with respect to the serosal (blood) side (7–10). The prime candidate for active ion transport has, therefore, been chloride. However, the fact that many studies have found that ouabain causes a reduction or abolition of the potential and short-circuit current suggests that sodium may also play a role, although the effects reported vary greatly. This variation is probably due to the unsatisfactory nature of the preparation, usually the iris–ciliary body, which requires that the pupil is blanked off by glueing a disc to its perimeter and that the serosal side is cut away from the underlying tissue, which causes much damage, provides a barrier to diffusion, and introduces unstirred layers.

In Sections 3 and 4 we describe how patch and whole-cell clamping can be applied to single, isolated ciliary epithelial cells, thereby overcoming some of the problems of the previous studies.

2. Lens epithelium

We have set out to measure the trans-epithelial resistance and potential of the lens epithelium, which is a single layer of cells situated at the anterior surface of the lens, without the complications introduced by the presence of the bulk of the lens tissue.

2.1 The 'double chamber' applied to the lens epithelium

In this section we describe how the double-chamber technique is applied to a core of lens tissue dissected from the anterior surface of the lens.

2.1.1 Description of dissection

We have applied the technique to the lenses of rat, rabbit, and bovine eyes. The bovine eyes were obtained from a local abattoir, and the best results were gained from tissue less than 2 h old, although we have obtained satisfactory results from material stored at 4 °C for up to 18 h. In addition, we have found that we can organ culture the lens in serum-free medium for several days and still record a potential and resistance within 20% of the fresh tissue values.

Protocol 1. Dissection of the lens

Materials
- artificial aqueous humour: NaCl, 125 mM; KCl, 4.4 mM; $NaHCO_3$, 10 mM; Hepes, 10 mM; $CaCl_2$, 2 mM; $MgCl_2$, 0.5 mM; glucose, 5 mM;

sucrose, 20 mM; adjust the pH to 7.4 with NaOH (1 M); osmolarity = 300 mOsm/l

Method

1. Make an incision in the eye at the corneal limbus (the line made by the junction of the cornea with the sclera) with a scalpel.
2. Using very sharp scissors, cut around the limbus and remove the cornea.
3. Keep the surface of the exposed lens moist with a buffered salt solution (artificial aqueous humour).
4. With a scalpel cut a central core of approximately 7 mm in the bovine, 5 mm in the rabbit, and 2 mm in the rat. This dimension is determined by the inner diameter of the holding rings (see *Figure 1*).
5. With fine scissors, cut away the surrounding tissue, leaving the core of tissue exposed.
6. Line the holding rings with Vaseline. Using fine scissors, undercut the core and lift away. Place the core of lens tissue into one holding ring (which should have a circular opening smaller than that of the core of tissue). A second holding ring slides over the first and the two then slide into the disc that divides the chamber.

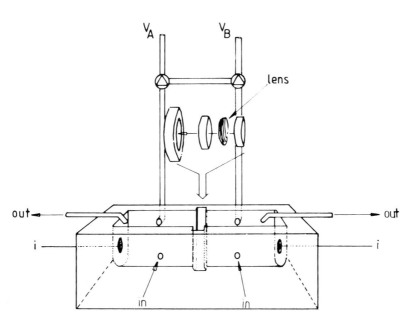

Figure 1. Diagram of lens holder assembly and double chamber. i, current electrodes. V_A and V_B are the voltage-recording electrodes for chambers A and B respectively. Solution is perfused in by gravity (in) and removed by suction (out).

2.1.2 Electrical set-up

The lens assembly is placed into a bath which separates the two surfaces of the lens tissue (*Figure 1*). There are many different ways of doing this, but we choose to have an open bath with a hemi-cylindrical well. The disc into which the lens is placed slides into a groove half way along this bath, thereby dividing it into two.

The voltage-recording electrodes are calomel electrodes (type CRL, Russell Electrodes). These are connected to tubes containing 3 M KCl, bridged in turn to saline by agar made up in 3 M KCl; the ends are separately equilibrated with 3 M KCl and saline, respectively. This minimizes junction potentials. Just before these saline tubes enter the two chambers of the bath, two three-way taps allow them to be connected; the electrodes can thereby be short-circuited. The voltage-recording electrodes are connected to a DC microvolt amplifier (Transducer Laboratories) with high gain, high input impedance, and low noise.

The end wall of each bath contains a silver/silver chloride disc electrode (8 mm) for current passage. This allows a symmetrical current flux through the chamber. These electrodes are connected to a battery-powered current source. This circuit is kept isolated from the voltage-recording circuitry.

The chamber and the electrodes are placed inside a Faraday cage to reduce electrical interference.

2.2 Verification of measurements

The trans-epithelial potential and resistance need to be verified since there are several sources of error associated with this technique.

2.2.1 Verifying potential measurements

With the lens in the chamber in the perfused bath maintained at 37 °C, high trans-epithelial potentials (V_{TE}) can be recorded, the positive voltage arising at the basolateral, aqueous-facing surface (*Table 1*). These were stable for up to 5 h.

The validity of these measurements was confirmed in the following ways:

(a) *Measurement of voltage and resistance before and after removal of epithelium.* This confirms that the potential and resistance recorded are not due to baseline shifts.

(b) *Measurement from posterior capsule.* Since the Na^+/K^+-ATPase activity is mainly restricted to the anterior surface of the lens, the posterior capsule should not be capable of generating a potential and therefore acts as a convenient control.

(c) *Amphotericin B.* This pore-forming antibiotic permeabilizes membranes. Therefore, depending on the side to which it is added, it should short-circuit the pump-generated trans-epithelial potential. When added to the apical side, it caused a small increase in V_{TE} which we suspect is due to the removal of a rate-limiting step in sodium transport, namely its entry into the cells across this membrane (Note: the pumps are situated on the basolateral membrane). When added to the basolateral side of the tissue, an immediate loss of potential occurred due to the short-circuiting of the pumps.

(d) *Temperature sensitivity.* If the potential is indeed generated by metabolically driven pumps (Na^+/K^+-ATPases), then it should be temperature sensitive. Diffusion potentials will be only very slightly temperature sensitive (0.085 mV/°C/decade of concentration difference, since $E_i = RT/z_iF.\ln\{C_i\}$, where E is the diffusion potential for ion i, R and F are the gas and Faraday's constant, respectively, z is valency, T is the temperature, and C is the ionic concentration gradient). We found a temperature dependence of 0.65 mV/°C in the rabbit and 0.26 mV/°C in the bovine.

(e) *Pump inhibitor.* If the trans-epithelial potential is indeed generated by the activity of the Na^+/K^+-ATPase then it should be abolished by inhibitors of the enzyme (e.g. ouabain). We found that about 75 and 80% of the potential was ouabain sensitive in the bovine and the rabbit lens epithelia, respectively.

Table 1. Trans-epithelial potential (V) and resistance (R)

	V (mV)	R ($\Omega.cm^2$)	n[b]
Rabbit lens epithelium	22.0 ± 0.6[a]	130.3 ± 17.1	18
Bovine lens epithelium	11.1 ± 0.4	116.6 ± 8.7	24

[a] Mean ± standard error. Data from Zhang and Jacob (11).
[b] n, number of experiments.

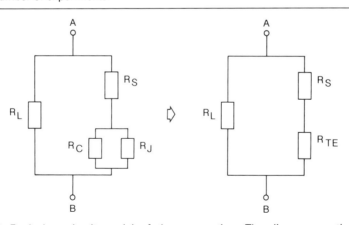

Figure 2. Equivalent circuit model of the preparation. The diagram on the left is simplified by 'lumping' the cell (R_C) and junctional (R_J) resistance to give R_{TE}, the trans-epithelial resistance (right-hand diagram). R_L, leak resistance; R_S, series resistance caused by remaining fibre cells; A, chamber A; B, chamber B.

2.3 Resistance measurements

Trans-epithelial resistance measurements are complicated by the possible existence of a parallel leak pathway around the edge of the tissue (see *Figure 2*). The following is an explanation of how to make educated guesses about the various resistances in the presence of such a leak.

The total resistance across a tissue (R_T) in a double chamber is given by the following expression:

$$1/R_T = 1/R_L + 1/(R_{TE} + R_S) \tag{1}$$

where R_L is the parallel leak resistance and R_{TE} is the trans-epithelial resistance, made up of a parallel combination of the cellular and paracellular pathways ($R_C \parallel R_J$), and R_S is the series resistance due to the remaining fibre cells.

In *Figure 2* the circuit is simplified (from left to right) by 'lumping' the parallel combination of the cell (R_C) and the junctional pathway (R_J) together as R_{TE} (trans-epithelial resistance). If the tissue is a 'tight' epithelium, (i.e. the cells are joined by high-resistance zonulae occludentes) then R_J will be large and thus R_{TE} will be large (R_C is high, $\geq$ 10 000 $\Omega.cm^2$). On the other hand, if the epithelium is 'leaky' then R_{TE} will be low, by definition R_L is low, and thus whatever the value of R_{TE}, R_T will be low. However, if there is no leak (i.e. no R_L), then R_T will depend on R_{TE} directly. A low measured R_T could indicate either a leaky epithelium or a tight epithelium with a leak pathway. The problem is therefore to distinguish a 'leaky' from a 'tight' epithelium in the presence of a leak pathway (R_L).

We measure R_T to be 240 $\Omega.cm^2$, which decreases to 120 $\Omega.cm^2$ after epithelium removal (i.e. removal of R_C and R_J in *Figure 2*), from which we can say that, if there is no leak, then $R_T = T_{TE}$ and $R_{TE} = 120\ \Omega.cm^2$. The presence of a leak would, of course, shunt the epithelium and hide the true value of its resistance.

In order to determine the relative value of R_L one must block (greatly reduce) the trans-epithelial conductance. During such a blocking experiment (with furosemide), we observed an increase in R_T from 240 to 390 $\Omega.cm^2$. Now this implies that the value of R_{TE} is the same order of magnitude as the leak: if R_{TE} was ten times larger than R_L it could only cause a change of approximately 10% to R_T at most; similarly, if it were a hundred times larger, then the most it could influence R_T would be by about 1%. In fact, if we assume that the trans-epithelial pathway becomes so resistive that most of the current flows through the leak, then $R_L \approx 390\ \Omega.cm^2$. From this it follows that $R_{TE} \approx 624\ \Omega.cm^2$. This value includes both the cellular and paracellular parallel combination ($R_C \parallel R_J$) in series with the resistance of the remaining fibre cells. Using this value of the leak one can calculate from the experimental data above that $R_S \approx 173\ \Omega.cm^2$, and, therefore, $R_C \parallel R_J \approx 450\ \Omega.cm^2$.

Thus, in the absence of a leak, $R_{TE} = 120\ \Omega.cm^2$, and if there is a leak then $R_{TE} \approx 450\ \Omega.cm^2$. In either case the epithelium falls into the category of a 'leaky' epithelium.

3. Ciliary epithelial cells

In this section we describe the patch and whole-cell clamping of single epithelial cells. There are two types of ciliary epithelial cells and both contribute to epithelial transport. They are coupled together to form a functional syncytium.

3.1 Preparation of fresh cells (pigmented and non-pigmented)

Protocol 2 describes the preparation of fresh cells, both pigmented and non-pigmented ciliary epithelial cells.

Protocol 2. Preparation of fresh cells

Materials
- low chloride solution (g/500 ml: sodium gluconate 15.26; KCl, 0.185; NaHCO$_3$, 0.21; glucose, 0.45; sucrose, 3.42; Hepes, 1.19; adjust the pH to pH 7.4)
- trypsin/EDTA (0.25% trypsin/0.02% EDTA in low chloride solution)

Method
1. One bovine eye is needed. Remove the cornea from the eye by making a circular incision at the limbus.
2. Remove the iris by cutting around its base at the irido-corneal angle with fine (preferably curved) scissors. This reveals the tips of the ciliary processes lying on the lens.
3. Cut the tips (1 mm) off the ciliary processes and sweep them up onto the surface of the lens. Use a wetted Pasteur pipette or 'filling tube' to suck up the ciliary process tips.
4. Remove the tips to a small tube containing a low chloride (Ca^{2+}-, Mg^{2+}-free) buffered salt sterilizing wash solution containing antibiotics: penicillin (100 IU/ml), streptomycin (100 µg/ml), gentamycin (100 µg/ml), amphotericin B (7.5 µg/ml). This solution sterilizes the preparation and prevents cell swelling.
5. Transfer to a large Petri dish containing 10 ml trypsin/EDTA.
6. Incubate at 37 °C for 17–20 min in a shaking water bath.

Protocol 2. *Continued*

7. Narrow the end of a Pasteur pipette in a flame. After wetting, transfer the tips to 10 ml low Cl^- solution with 10% fetal calf serum (FCS) to inhibit the trypsin.
8. Triturate 2–3 times to dissociate single cells.
9. Allow to settle for 3–5 min. Remove the tips with a Pasteur pipette.
10. Centrifuge the supernatant for 5 min at 500 g to pellet the cells.
11. Discard the supernatant and resuspend the pellet in 10 ml Cl-free solution.
12. Repeat steps 10 and 11, finally resuspending in 1 ml of medium (E199) with 10% FCS. Add 120 µl per 13 mm cover slip.
13. Leave overnight to allow the cells to attach.

Figures 3 and *4* illustrate freshly prepared cells using *Protocol 2* as seen by scanning electron microscopy.

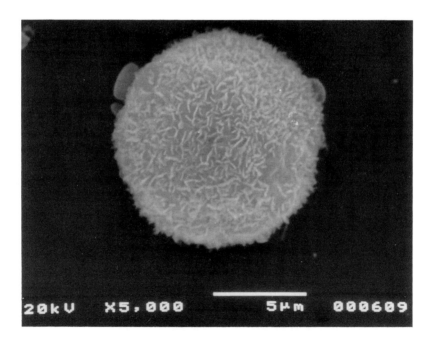

Figure 3. A scanning electron micrograph of a bovine pigmented ciliary epithelial cell freshly isolated on a Percol gradient. Note the profusion of microvilli. (Figure courtesy of Jane Morgan.)

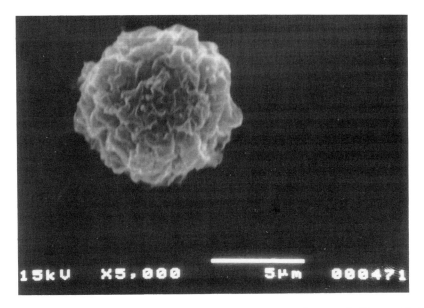

Figure 4. A scanning electron micrograph of a bovine non-pigmented ciliary epithelial cell. The cells are smaller than the pigmented cells and membranous folds replace the microvilli (Figure courtesy of Jane Morgan.)

3.2 Preparation of cells for culture

Stelling and Jacob (12) have shown that cultured bovine pigmented ciliary epithelial cells lose an inward rectifier K^+-current in culture (see Section 5). In view of this and other reports of changes occurring to cultured cells, for example changes to the kinetics of the Na^+/K^+-ATPase (13), one has to be a judicious in the use of cultured cells.

3.2.1 Pigmented cells

It is much easier to grow pigmented than non-pigmented ciliary epithelial cells. In a mixed culture the pigmented cells take over in a very short space of time because of their more rapid proliferation rate.

Protocol 3. Preparation of pigmented ciliary epithelial cells

Materials
- DPBS: Dulbecco's phosphate-buffered saline
- FCS: fetal calf serum

Protocol 3. *Continued*

Method

The following protocol is designed for four bovine eyes.

1–3. Carry out steps 1–3 in *Protocol 2*.

   4. Remove the tips to a small tube containing a buffered salt sterilizing wash solution, e.g. DPBS containing antibiotics: penicillin (100 IU/ml), streptomycin (100 μg/ml), gentamycin (100 μg/ml), amphotericin B (7.5 μg/ml).

   5. Transfer to a large Petri dish containing 10 ml of 0.25% trypsin/0.02% EDTA in DPBS (or a similar buffered saline).

   6. Incubate at 37 °C for 20–25 min in a shaking water bath.

   7. Narrow the end of a Pasteur pipette in a flame. After wetting the pipette, triturate the tips 2–3 times in trypsin to remove blood cells. Discard the trypsin and keep the tips.

   8. Add approximately 10 ml of medium with FCS to neutralize the remaining trypsin.

   9. Triturate to remove ciliary epithelial cells, taking care not to form air bubbles which will break up the cells.

   10. Allow the tips to settle. Discard the tips.

   11. Centrifuge for 5 min at 500 g to pellet the cells.

   12. Discard the supernatant and resuspend the pellet in 10 ml DPBS.

   13. Repeat steps 11 and 12, finally resuspending in 8–10 ml of culture medium.

   14. Transfer 2 ml of the resuspended cells to each 35 mm Petri dish.

3.2.2 Non-pigmented cells

In order to obtain a culture of pure non-pigmented ciliary epithelial cells, it is necessary to separate them from the pigmented cells. This is achieved by use of a discontinuous Percol gradient (*Protocol 4*).

Protocol 4. Percol gradient preparation of non-pigmented ciliary epithelial cells for culture

Materials

- 45% Percol solution (4.5 parts Percol (Pharmacia), 0.5 parts × 10 DPBS (Gibco), 5.0 parts × 1 DPBS)

Method

Six bovine eyes are optimal.

1–11. Carry out steps 1–11 in *Protocol 2*.

12. Resuspend the pellet in 6 ml of a buffered salt solution (e.g. low-Cl⁻ solution (see *Protocol 2*)).

13. Put 6 ml 45% Percol solution into a 15 ml centrifuge tube. Add this solution directly to the tube bottom to prevent it sticking to the sides.

14. Mark the position of the Percol meniscus. Carefully layer the cell suspension over the Percol.

15. Centrifuge for 5 min at 800 g in a swing-out rotor.

16. The non-pigmented cells band at the interface. The pigmented cells collect in a pellet at the bottom.

17. Wash the non-pigmented cells several times by pelleting and resuspending, finally resuspending in serum-free culture medium (we use Ames' Medium).

18. Plate out in serum-free medium. Allow the cells to adhere (1–2 days).

19. Re-feed with a medium containing 10% FCS.

3.3 Preparation of electrodes (pipettes)

Patch and whole-cell electrodes, referred to as pipettes, are best pulled with a two-stage puller, but can be pulled using a single-stage puller (with some practice).

3.3.1 Pulling, breaking back, fire-polishing

This section describes the preparation of electrodes for patch and whole-cell recording. There are a great range of different glasses that can be used to pull pipettes; borosilicate (soft, good sealing properties, moderately noisy), aluminosilicate (hard and therefore requires high temperatures to pull, low noise), soda glass (soft, noisy; see reference 14). A good range of glasses can be obtained from Clark Electromedical. Patch-clamping has slightly different requirements, for example electrodes need to have a higher resistance (8–20 MΩ) and do not need to have a rapid taper. For whole-cell recording the objective is to pull an electrode (2–6 MΩ) that tapers very rapidly to a smoothly rounded end with a circular opening. Such a conformation allows a complete and rapid exchange of the pipette contents with the solution inside the cell and is achieved with a two-stage puller (e.g. Model P-30, Sutter Instrument Co.). Because of the low resistance of the electrodes, either filamented or non-filamented glass can be used (but most people seem to use filamented glass). Once pulled, the filamented electrodes are back-filled by placing a drop of filling solution at the top and allowing capillary action to take the solution to the tip. Non-filamented electrodes have to be filled by suction.

3.3.2 Electrode (pipette) size test
Having pulled the electrode as described in Section 3.3.1, attach it to a 5 ml or 10 ml syringe with silicone or polythene tubing. The pressure required (in millilitres) to cause bubbling under methanol is proportional to the size of the tip. Once a calibration curve has been constructed (measuring the electrode resistance with the patch-clamp), it is possible to screen new electrodes very quickly.

3.3.3 Making electrodes with a single-stage puller
It is possible to make suitable electrodes with a single-stage puller, but the electrodes need to be broken back and fire-polished. The electrode can be broken back on the stage of a microscope with a slide holder/positioner. Attach the electrode to a slide with plasticine. Mount a glass rod with an optically smooth end (best obtained by scoring the glass rod with the glass cutter and snapping) so that the end is in the centre of the field of view. Move the electrode towards the glass rod until the reflected image and the electrode meet. Tap the microscope. With practice it is possible to make usable electrodes in this way. Subsequent fire-polishing improves the success rate. A device can be rigged up using a micromanipulator and a small heating coil coated with glass to prevent deposition of metal on the electrode tip.

3.3.4 Sylgard coating
Coating the electrode with Sylgard (Down Corning elastomer 184) to within 50–100 μm of the tip greatly improves the noise and reduces the capacitance of the chamber/electrode combination. The Sylgard can be made up beforehand, aliquotted and stored at −20 °C. It can be applied to the electrode with a fine paintbrush or a mounted needle and then cured (hardened) by heating. To do this, the coated electrode is placed in the heating coil of the electrode puller with gentle heat until a puff of smoke is observed. The Sylgard will then be firm to the touch, but note that excessive heat will cause it to craze.

3.3.5 Filling
Electrodes for whole-cell recording are filled with a solution similar in ionic composition to the inside of the cell. The contents of the pipette dialyse the cytoplasm within a few minutes and many substances are washed out, so the effect of this needs to be considered. In common with many other people, we add ATP and Mg^{2+} to the pipette and in some experiments include cyclic nucleotides which are known to modulate ion channels.

Table 2. Pipette filling solution

	Concentration (mM)	Final [Ca^{2+}] (M)
KCl	56	
K gluconate	84	
Hepes	10	
MgCl$_2$	2	
Sucrose	20	
EGTA	1.1	
CaCl$_2$	0.1	10^{-8}
	0.43	5×10^{-8}
	0.55	10^{-7}

Adjust the pH to 7.25 with 1 M NaOH ($\approx$ 6 mM Na$^+$) and then filter through a 0.2 μm filter before use.

The solution in *Table 2* was designed to separate the reversal potentials of the major ions. Thus: $E_{Na} = +78$ mV, $E_K = -84$ mV, $E_{Cl} = -20$ mV and $E_{cation} = 0$ mV (if this solution is used with a conventional saline, at room temperature and with 6 mM Na$^+$ in the pipette). It has a low calcium-buffering capacity. Thus, if it is required that the [Ca^{2+}]$_i$ is clamped, then a higher concentration of EGTA or some other Ca^{2+} buffer should be used. Evans and Marty (15) found that it was necessary to use 40 mM HEDTA to overcome the intrinsic buffering capacity of the cell.

The osmolarity of the intracellular and extracellular solutions should be carefully matched. Care must be taken to control for junction potentials that develop at the tip of the electrode and are compensated for by the zeroing procedure prior to seal formation, only to disappear once a gigaohm seal has been formed and the electrode is no longer in contact with the bath solution (16).

4. Recording from epithelial cells using the patch-clamp

In this section we discuss the use of the patch-clamp to look at single channels and whole-cell currents. Several excellent books have been written on the practical aspects of single channel recording and there is not the space to attempt a detailed description here. *The Plymouth Workshop Handbook* edited by Standen, Gray, and Whitaker (17) is a good introduction to the techniques of patch-clamping and the analysis and interpretation of data. *Single Channel Recording* edited by Sakmann and Neher (18) gives a very detailed account of the subject.

4.1 Patch recording

Single channel recording from pigmented ciliary epithelial cells.

4.1.1 Sealing onto the cell

The presence of numerous microvilli and membranous folds (see *Figures 3* and *4*) makes obtaining a gigaohm seal far more difficult than, for example, with cultured neurones. It is remarkable that it is possible at all. *Protocol 5* describes how to obtain a high-resistance seal onto the cell surface prior to recording single channel activity.

There are a number of reasons why a seal may not be formed:

- bad electrode profile
- dirty glass
- dirty solutions
- cells need cleaning enzymatically
- cells too flat (may need to round-up)
- bad luck

Protocol 5. Forming a gigaohm seal to a cell

1. After preparing and filling the patch pipette ($R_e = 8-20$ MΩ, see Section 3.3) insert into pipette holder (with air-tight seals and suction line).
2. Apply positive pressure (0.1–0.5 ml from a syringe attached via the suction line to the pipette holder).
3. When the pipette is in the bath, switch on the resistance test pulse (most patch-clamps have a small, continuous voltage pulse for this purpose).
4. Zero the baseline (but see reference 16 for a description of the erroneous junction potential compensation that this introduces).
5. Move the electrode to the cell surface. When it touches (you will see refractive index changes, a change in the baseline current and/or an increase in resistance) release the pressure. The resistance may increase still further.
6. Apply suction (0.1–1.0 ml is usually adequate). The resistance should increase to $\geq 10^9$ Ω (a gigaohm seal). This may happen immediately or over some minutes, in which case the application/release of suction a number of times may gradually form the high-resistance seal required. A solution containing 10 mM Ca^{2+} aids seal formation. If used, this should be washed out immediately afterwards.
7. Compensate for the capacitance of the bath/pipette/patch combination (Sylgard coating the tip of the pipette will help).
8. Switch off the resistance test pulse.
9. Increase the gain on the patch-clamp amplifier to observe single channel currents. It may be necessary to alter the pipette voltage (a negative

pipette potential depolarizes a cell-attached patch). Apply suction/ agonist or transmitter to activate channels.
10. To excise the patch to give an outside-out configuration, simply pull back the pipette into a Ca^{2+}-free bathing solution. A vesicle may form (detected by the 'round-shouldered' appearance of the single channel events) which can be ruptured by air exposure.

4.1.2 Cell-attached recording

It has always been assumed that epithelial cells plate down and gradually become polarized such that their apical membrane is uppermost. Evidence from scanning and transmission microscopy confirm this notion. This would allow the apical membrane to be explored with the patch electrode and the allocation of certain channels to a particular membrane to be made. However, recent immunocytochemical evidence has shown that this is not necessarily the case. Labelled antibodies to a surface marker, the apical Na^+ channel, were found to bind to both membranes in cultured A6 cells (19). The assigment of a particular channel to one or other surface of cultured cells cannot, therefore, be made on the basis of single channel studies alone.

Figure 5 shows single channel activity recorded from a cultured pigmented ciliary epithelial cell in the cell-attached recording mode. The membrane patch was depolarized by 50 mV to activate the channel and on the right the probability that a channel was open is plotted as a function of pipette potential (note that a negative pipette potential represents membrane patch depolarization). This channel is a Ca^{2+}-activated K^+ channel (conductance = 182.1 ± 7.0 pS, $n = 4$).

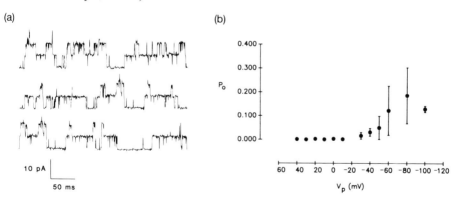

Figure 5. (a) Single channel activity recorded from a cell-attached patch of a pigmented cell, $V_{pipette} = -50$ mV; pipette solution contained 140 mM K^+. (b) Probability of a channel being open increases as a function of membrane depolarization. Cell-attached patch, 140 mM K^+ in pipette. Note that a negative pipette voltage represents a membrane depolarization. Data from four separate cells (mean ± SEM). P_o, open probability; V_p, pipette potential.

4.1.3 Excised inside-out recording

Switch on the resistance test pulse, set the pipette voltage to zero. With the electrode in the cell-attached mode, pull back the pipette into a calcium-free bathing solution with the desired ionic concentrations (usually high K^+ to mimic the intracellular environment). If the current baseline jumps off scale you have probably lost the patch. If not, and you still have a gigaohm seal, then switch off the test pulse and you may see channel activity, or applying a pipette voltage may activate channels.

4.1.4 Excised outside-out recording

After obtaining the whole-cell recording mode (see Section 4.2), pull back the electrode (a little suction may be tried). The capacitative spike will decrease as will the time constant.

4.2 Whole-cell recording

This technique is less well described in the text books and careful reading of the appropriate patch-clamp manual is generally the best source of information.

Protocol 6. Whole-cell recording

1–6. Carry out steps 1–6 in *Protocol 5*.

  7. Switch in the desired holding potential (in the whole-cell configuration negative pipette potentials give negative membrane potentials).

  8. Fix the position of the current response to the voltage test pulse in the centre of a storage oscilloscope screen to get a good view of the capacitative spike, and balance it out using the capacity compensation (we can never remove all of this in the pigmented cells).

  9. Dial in the series resistance compensation until the current rings (oscillates) and then turn it back until it just stops. During this process the time constant for the decay of the capacitative spike should decrease.

  10. Apply suction until the capacitative spike and the decay time constant suddenly increase. This ruptures the patch of membrane under the pipette. The increased capacitative current reflects the increased membrane area exposed by breaking through.

  11. Turn off the test pulse.

  12. Start voltage-clamp pulse protocols to monitor cell currents.

  13. Allow some minutes ($\approx$ 3 min) for the pipette contents to dialyse the cell interior.

4.2.1 Suction technique

A brief outline of this technique is given in *Protocol 6*. A gigaohm seal is achieved as for cell-attached patch recording (*Protocol 5*) and then the patch of membrane beneath the pipette is either ruptured by suction or permeabilized with a pore-forming antibiotic such as nystatin (see Section 4.2.2). The use of fresh cells allows an even space clamp since cells are spherical. Flattened cells or cells with long dendrites or processes cannot be evenly or completely voltage-clamped and this must be accounted for (see reference 20). Using cultured cells it is necessary to re-perfuse with trypsin to round-up the cells to avoid this problem, but one then must confirm that this does not change cell properties. We have found that perfusing a pigmented ciliary epithelial cell with 0.25% trypsin for 10 min has no effect on the whole-cell currents. Rupturing the membrane patch beneath the pipette allows the contents of the pipette to dialyse the cell. This may wash out important cytosolic constituents. This can be avoided by using the permeabilized patch technique (Section 4.2.2).

It is often useful to know the cell capacitance (as an index of cell size). The expression of the currents per unit capacitance allows cells of different sizes to be compared. In addition, the relative sizes of the series and membrane resistance need to be considered. In some cells the membrane resistance is low (especially when many channels are activated, e.g. during an action potential) and may even be of the same order of magnitude as the series resistance. To avoid the problem of a voltage divider (the command voltage being split across the series resistance and the membrane), series resistance needs to be compensated for, as follows.

4.2.2 Calculating cell capacitance and series resistance from whole-cell voltage-clamp currents

(a) Select a current trace with no active currents of a cell for which the value of the series resistance compensation (R_{comp}) is known.

(b) Fit an exponential to the capacitative current decay to obtain the current at $t = 0$ (i_0), the time constant (τ), and the steady stage current (i_{ss}).[a]

(c) The value of the y-axis intercept gives the current, i_0, through the series resistance (pipette and cytosolic resistance, R_S). Thence $R_S = V/i_0$, where V is the voltage pulse.

(d) The capacitance, $C = \tau/R_S$.

(e) The percent of series resistance compensated for is given by $R_{comp}/(R_S + R_{comp}) \times 100$. It is not usually more than about 60%.

(f) If the uncompensated R_S is 100-fold larger than R_m (membrane resistance), then the voltage divider ratio is 1:99 and a 1% error is introduced into a clamped potential. A less favourable ratio (i.e. smaller difference) requires the necessary correction to the voltage to be made.

[a] Note that filtering and sampling frequency will affect the height of the capacitative spike.

4.2.3 Permeabilized-patch technique

A pore-forming antibiotic is used to render the patch beneath the pipette permeable, thereby giving a low resistance access to the cell interior. The small monovalent cations equilibrate across the permeabilized membrane very rapidly, but there is some confusion as to whether the same happens for Cl^-, and almost certainly the same does not happen for Ca^{2+}. Refer to Horn and Marty (21) for a detailed account. While this technique causes minimal disturbance to the intracellular environment, and this is its major advantage, it has the drawbacks that $[Ca^{2+}]_i$ control is difficult, if not impossible, and that junction potentials across the permeabilized patch of unknown value may develop. These have to be carefully controlled for.

We have attempted this technique with nystatin and amphotericin B in both pigmented and non-pigmented cells. Only with amphotericin B have we been successful. The whole-cell currents so obtained are the same as those obtained with the suction technique (see Section 4.2.1).

4.3 Data analysis

Two software programs for data acquisition, recording, and analysis have been developed in the UK. One of these, a commercial package from Cambridge Electronic Design, is quite expensive but does provide a very comprehensive (regularly updated) suite of programs for pulse generation, single channel and whole-cell recording, and analysis. The other is available free from Dr John Dempster, University of Strathclyde and is a very competitive product. Both can use the CED 1401 laboratory interface that is becoming a standard in the UK. American- and French-produced software, requiring specific (different) hardware, is available, but problems with after-sales service and technical back-up can make otherwise good products less favourable.

5. Differences between fresh and cultured cells

5.1 Size difference

Cultured pigmented ciliary epithelial cells increase in size with time in culture. This is reflected in the cell capacitance (C) measurements. Fresh cells have a capacitance of $C = 11.5$ pF which increases from 13 pF at Day 8, to 22 pF at Day 13, and to 32 pF at Day 23.

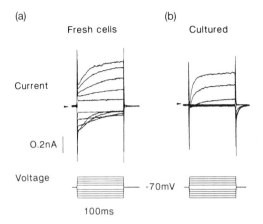

Figure 6. Whole-cell currents recorded from (a) fresh and (b) cultured pigmented ciliary epithelial cells in response to a series of hyperpolarizing and depolarizing steps (from 20 to 100 mV in 20 mV steps) from a holding potential of −70 mV. The inward current activated by hyperpolarization disappears in culture.

5.2 Disappearance of inward rectifier

Figure 6 shows the whole-cell currents from fresh cells compared with cultured cells in response to a series of hyperpolarizing and depolarizing voltage steps from a holding potential of −70 mV. In fresh and cultured cells, depolarizing voltage steps activate a large, delayed outward current which has been shown to include a voltage-activated K^+ current and a Ca^{2+}-dependent K^+ current (22). In freshly prepared cells, hyperpolarization activates an inward current that has been shown to be the inward rectifier K^+ current (12). After 2 days in culture the inward rectifier current begins to diminish and it disappears completely after 8 days.

References

1. Kinsey, V. E. and Reddy, V. N. (1965). *Invest Ophthalmol. Vis. Sci.*, **4**, 104.
2. Davson, H. (1990). *Physiology of the Eye*, Chap 4, pp. 139–202. Macmillan Press, London.
3. Duncan, G. (1969). *Exp. Eye Res.*, **8**, 406.
4. Jacob, T. J. C. and Duncan, G. (1981). *Nature*, **290**, 704.
5. Harris, J. E. and Gersitz, L. B. (1951). *Am. J. Ophthalmol.*, **34**, 131.
6. Harris, J. E. and Gruber, L. (1962). *Exp. Eye Res.*, **1**, 372.
7. Holland, M. G. and Stockwell, M. (1967). *Invest. Ophthalmol. Vis. Sci.*, **6**, 401.
8. Watanabe, T. and Saito, Y. (1978). *Exp. Eye Res.*, **27**, 215.
9. Kishida, K., Sasabe, T., Manabe, R., and Otori, T. (1981). *Jpn. J. Ophthalmol.*, **25**, 407.

10. Burstein, N. L., Fischbarg, J., Liebovitch, L., and Cole, D. F. (1984). *Exp. Eye Res.*, **39**, 771.
11. Zhang, J. J. and Jacob, T. J. C. (1992). *J. Physiol.*, **452**, 58P.
12. Stelling, J. W. and Jacob, T. J. C, (1991). *J. Physiol.*, **435**, 91P.
13. Whikehart, D. R., Montgomery, B., and Hafer, L. M. (1987). *Curr. Eye Res.*, **6**, 709.
14. Rae, J. L. and Levis, R. A. (1984). *Biophys. J.*, **45**, 144.
15. Evans, M. G. and Marty, A. (1986). *J. Physiol.*, **378**, 437–60.
16. Barry, P. H. and Lynch, J. W. (1991). *J. Membr. Biol.*, **121**, 101.
17. Standen, N. B., Gray, P. T. A., and Whitaker, M. J. (ed.) (1987). *Microelectrode Techniques; the Plymouth Workshop Handbook*. The Company of Biologists, Cambridge.
18. Sakmann, B. and Neher, E. (ed.) (1983). *Single Channel Recording*. Plenum, New York.
19. Kleyman, T. R., Kraehenbuhl, J.-P., and Ernst, S. A. (1991). *J. Biol. Chem.*, **266**, 3907.
20. Rall, W. and Segev, I. (1985). In *Voltage and Patch-Clamping with Microelectrodes* (ed. T. G. Smith, H. Lecar, S. J. Redman, and P. W. Gage), pp. 191–215. American Physiological Society, Bethesda.
21. Horn, R. and Marty, A. (1988). *J. Gen. Physiol.*, **92**, 145.
22. Jacob, T. J. C. (1991). *Am. J. Physiol.*, **261**, C1055.

3

Isolation of sinoatrial node cells

H. F. BROWN

1. Introduction

The dissociation of cardiac tissue into single cells opened up new possibilities for physiological investigation which have led in the last decade to major advances in knowledge of heart function at the cellular level. Cardiac cell isolation was pioneered in the 1970s (1) and was quickly adopted by many other groups. The basic principle is simple: cardiac tissue is exposed to enzymes (always collagenase and often other enzymes such as trypsin, protease or elastase) in a Tyrode solution containing a very low concentration of calcium. Under these conditions, the cell junctions are disrupted and will seal over to give intact but separate cells. In practice, it is not always easy to obtain cells that remain calcium tolerant when returned to physiological levels of calcium and that are neither too fragile nor too 'tough' for procedures such as patch-clamp recording, while retaining their normal membrane pumps and receptors. Nevertheless, the isolation of ventricular and atrial cells is now successfully carried out in many laboratories.

2. Separation of sinoatrial node cells

2.1 General introduction

The sinoatrial (SA) node, although the first region of the heart to be excited at each beat, has always been the last to which investigative methods have been successfully applied. Cell separation has been no exception to this general rule: the difficulties have arisen both from the small size of the nodal cells and from their dense packing in connective tissue. Thus it is difficult to give nodal tissue enough enzyme treatment to liberate the cells from the surrounding connective tissue, while leaving them calcium tolerant and with their membrane channels and receptors intact.

2.2 Appearance of healthy SA node cells

Healthy isolated SA node cells appear spindle shaped (*Figure 1*), with a clear membrane. When superfused with Tyrode solution at 37 °C they are spontaneously active, giving visible beats, and continue beating for 1–2 h. *Figure 2A* is a diagram of the SA node showing the region from which strips of tissue are cut for cell separation. Patch-electrode recording in the whole-cell mode will record the spontaneous pacemaker activity from the cells (*Figure 2B*).

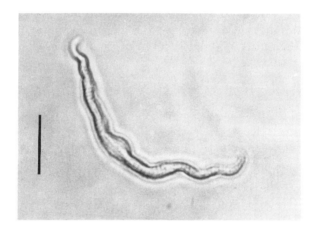

Figure 1. An isolated cell from the sinoatrial node of the rabbit. Bar represents 25 μm.

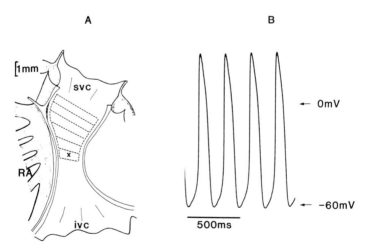

Figure 2. (A) Diagram of the SA node region of the rabbit. The right atrium (RA) and superior vena cava (svc) have been cut open. The dotted lines indicate the tissue removed for cell isolation. x, central region of the SA node, where firing normally starts; ivc, inferior vena cava. (Adapted from a drawing by Dr Junko Kimura.) (B) Spontaneous electrical activity recorded from a single SA node cell with a patch electrode in the whole-cell configuration.

2.3 Methods of isolating SA node cells

2.3.1 A recommended method

The protocol which follows (*Protocol 1*) is one which *has* given healthy, vigorous isolated cells, but it is not guaranteed to do so! There seems to be a consensus, among those who have tried SA node cell isolation, that a procedure that is successful for a time then deteriorates, and no longer gives good cells. This usually coincides with a change in the batch of enzyme used. The protocol used in Oxford will first be given, followed by a survey of the variations favoured by other laboratories.

Protocol 1. Separation of SA node cells

Materials

- pure water: glass-distilled and then de-ionized (Milli-Q)
- components for Tyrode solution and for 'KB' solution (see *Table 1*), including 1 M stock solutions of $CaCl_2$ and $MgCl_2$
- collagenase (Worthington) and elastase (Sigma Type IIA)
- dissection dishes (with a layer of Sylgard for pinning out tissue) and instruments
- binocular microscope
- solid heating block set at 37 °C
- two small glass containers with pierced lids to take oxygenation tubes
- oxygen
- glass pipettes, stopwatch etc.
- rabbit (albino; 700–900 g)

Method

1. Make up the solutions shown in *Table 1*.
2. Take a tube of KB solution out of the freezer to thaw, before starting the separation procedure.
3. Make up 1 litre of Tyrode solution without adding any calcium. Add 15 µl of 1 M $CaCl_2$ solution. Shake.
4. To 500 ml of this, add $CaCl_2$ (885 µl) to make normal Tyrode solution. Warm this in the waterbath at 37 °C and oxygenate.
5. To 100 ml of the 15 µM Ca^{2+}-Tyrode add 300 µl of 1 M $MgCl_2$. To 10 ml of this solution, add 5 µl of sodium EGTA solution. Pour some (~ 3 ml) of this into a small glass container. Put this into the heating block at 37 °C to warm (EGTA solution, A).

Isolation of sinoatrial node cells

Protocol 1. *Continued*

6. Take 2 ml of the 15 µl Ca^{2+}-Tyrode and put in a small glass container in the heating block (Solution B).
7. Weigh out collagenase and elastase in a quantity sufficient to give 85 IU/ml collagenase and 23 IU/ml elastase, when dissolved in 2 ml (this will be about 1 mg of each). Keep in the refrigerator until it is time to use.
8. Kill a rabbit by cervical dislocation and quickly remove the heart. Place in the oxygenated, warmed normal Tyrode solution. Remove the SA node region; cut open the superior vena cava and pin the nodal region out in a small Sylgard-lined dish.
9. Cut out from the SA node an area about 7 × 3 mm; subdivide into two groups of three strips by cutting with a razor blade tool or a new scalpel blade (*Figure 2A*).
10. Place the strips in Solution A and leave them 5 min in this, in the heating block, bubbling the solution with oxygen.
11. Dissolve the preweighed enzymes in the 2 ml of Solution B. At the end of the 5 min in Solution A, transfer the strips to Solution B. Leave the two groups of strips of tissue in enzyme in the heating block (with oxygenation) for 10 min and 12 min.
12. After 10 and 12 min, transfer the tissue to 'KB' solution in the refrigerator.
13. Leave the tissue at least 1 h in KB solution at 5 °C.
14. With fine forceps, gently tease apart a strip of the tissue in a drop of KB medium to release the cells. This can be done in a small Petri dish under a binocular microscope and then the cells transferred, or it can be done directly in the perfusion bath.
15. Allow the cells 5 min to settle on the glass base of the perfusion dish, then perfuse first with Ca^{2+}-free Tyrode solution (1 min) and then with normal Tyrode solution warmed to 37 °C. Healthy, Ca^{2+}-tolerant cells will appear elongated in form, with a clear membrane (*Figure 1*); they will beat spontaneously.

2.3.2 Survey of methods of isolating SA node cells

Published methods for separating SA node cells show considerable differences. Reading 'between the lines', no method which reliably gives a high yield of healthy, spontaneously beating cells has yet been devised, although many of the methods described give some cells that show spontaneous electrical activity and are in a reasonably normal state. The methods have been developed from those used to separate ventricular tissue into individual cells,

Table 1. Composition of Tyrode and 'KB' storage solutions (12)

Normal Tyrode	Concentration (mM)
NaCl	140
KCl	5.4
$MgCl_2$	0.5
Glucose	10
Hepes	5
pH adjusted to 7.4 (at 37 °C) with 4 M NaOH	
KB Storage solution	
KCl	70
K_2ATP	5
$MgCl_2$	5
K glutamic acid	5
Taurine	20
Trisphosphocreatine	5
EGTA	0.04
Succinic acid	5
KH_2PO_4	20
Glucose	10
Hepes	5
pH adjusted to 7.0 with 1 M KOH solution	

which themselves started to give reliable cell separations about 10 years ago (2).

The one thing that all the methods have in common is a period of incubation in collagenase, but the concentration used and mode and duration of exposure of the SA tissue to this enzyme varies widely, as do the amounts of other enzymes used in combination with collagenase.

Several published methods for SA cell separation start with a Langendorff perfusion of the whole heart, first with Ca^{2+}-free Tyrode solution alone and then with collagenase included. A difficulty of continuing such a perfusion too long is, however, that collagenase so softens and distorts the tissue that it becomes hard to identify and isolate the SA node. One of the first published accounts of SA node cell separation (3) used a Langendorff perfusion with 0.4 mg/ml collagenase for 1 h. Photographed in the high-K^+ storage solution in which they were separated out from small pieces of nodal tissue, the isolated cells appeared rod-like rather than spindle shaped. Furthermore, most of the cells rounded up on exposure to normal (1.8 mM Ca^{2+}) Tyrode solution, although some rounded cells still showed spontaneous pacemaker activity.

This method of separation was employed by other investigators, and the rounded cells it produced were used for whole-cell voltage-clamp experiments and for single channel recording (4, 5). The Japanese groups who study isolated SA node cells have continued with fairly long exposures of SA node tissue (30 min or more) to low concentrations of collagenase, although more recently they have incubated strips of SA node tissue rather than using a Langendorff perfusion. Collagenase treatment is followed by a period in high-K^+ 'storage' solution. The resulting cells always round up on re-exposure to normal (1.8 mM Ca^{2+}), but have nevertheless yielded many results (e.g. 6, 7). An assessment of 'rounding up' is given below.

DiFrancesco et al. (8) introduced SA node cell separation methods consisting only of the enzyme incubation of excised tissue, with no prior Langendorff perfusion. After a short period in nominally Ca^{2+}-free, low-Mg^{2+} Tyrode solution containing 50 mM taurine and of pH 6.9, strips of nodal tissue were incubated for 15–30 min in a similar solution containing the enzymes collagenase and elastase, and 1 mg/ml albumin. The strips were moved around within the solution during this incubation by a special agitating device. The tissue strips were then moved to a high-K^+ (110 mM) solution (containing low Ca^{2+} and various 'nutritive' compounds) for a short period of further trituration to disperse the cells. These were then left for a further period in high-K^+ solution before return to normal Tyrode solution via a series of gradual increments in Ca^{2+} concentration. More recently, DiFrancesco and Tromba (9) have used a 0.2 mM Ca^{2+} enzyme solution, rather than 'Ca^{2+}-free' solution. DiFrancesco et al. (8) describe three types of cells: rod-like cells with clear striations, round cells, and 'spider cells' which often have one or more thin prolongations. It was this last type that they found most suitable for whole-cell patch electrode study of the hyperpolarization-activated inward current, I_f.

Brown et al. (10) found that the incubation of excised SA node tissue in a (Ca^{2+}-free) collagenase/elastase Tyrode solution with the minimum disturbance for 90 min gave single cells which were typically of a thin, elongated spindle shape, 10 μm in diameter, and 100 μm (range 70–170 μm) long (for further details see reference 11). It was subsequently found that cells which were the same elongated shape but electrophysiologically much livelier, showing spontaneous activity and good membrane currents under voltage-clamp conditions, could be obtained by drastically reducing the time of enzyme incubation to 10–15 min and following it by a period in high-K^+ 'storage' solution (12). It is this modified method which is given in *Protocol 1*.

Doerr et al. (13) used a high-K^+ (110 mM) solution for enzyme incubation of SA nodal strips, similar to the 'KB' storage solution in *Protocol 1*. Strips of SA node tissue were agitated in this for 40 min and then, after a wash in Ca^{2+}-free solution, were shaken in a tissue culture medium for 20 min to liberate single cells. The cells obtained were fairly elongated (dimensions 40 × 15 μm) and showed spontaneous activity.

Nathan (14) incubated pieces of SA node tissue in collagenase and DNase to separate cells which were then maintained in short-term culture. He describes two distinct types of cell, round and elongated, with different electrophysiological properties.

Van Ginneken and Giles (15) and Yatani and Brown (16) have used a combination technique: a Langendorff perfusion is followed by further incubation of excised SA node tissue in collagenase. Van Ginneken and Giles (15) obtained cells of several morphological types; cells that showed spontaneous activity were all of elongated form. They also found large (20 μm diameter) spherical cells, which were quiescent, and smaller (5–8 μm) spherical cells from which they could not record. Yatani and Brown (16) report cell dimensions of 30–60 μm by 8–10 μm. They do not, however, mention (or show) any spontaneous activity from their cells, which they studied at 20–22 °C—at which temperature spontaneous activity would be slow or absent. Both these studies were chiefly of I_f, which is very resistant to the run-down that affects other membrane currents (see below).

3. The shape of sinoatrial node cells: spindles, spheres, or spiders?

It seems that the true natural shape of the SA node cell is a thin, elongated spindle shape (4, 11) of dimensions about 10 × 100 μm, as shown in *Figures 1* and *4A*. Although there is evidence both from electron microscopy (17) and from light microscopy (author's own observations; reference 18) that there is a gradation in cell length from the centre to the periphery of the SA node, isolated cells that are substantially shorter than the dimensions give here have probably become partially rounded up in the separation process. Rounding up is something that SA node cells seem to do very readily when isolated, possibly because they contain rather few myofibrils and other intracellular structures. In the intact node, the cells will be supported by cell-to-cell contacts and by the substantial amounts of connective tissue present. Rounding up can occur without apparently greatly affecting the electrophysiological properties of the cells, and several studies have been carried out on such cells (4–7). It must, however, be borne in mind that fibroblasts, which are very abundant in the SA node region, closely resemble SA cardiomyocytes when both types of cells have rounded up on cell separation (19). If SA node cells are isolated and then cultured, they quickly round up (*Figure 3*), but these rounded cells will continue vigorous spontaneous beating for more than a week of culture (author's own observations).

Staining of isolated SA node cells shows that the typical cell has only a single nucleus (*Figure 4A*). The presence of more than one nucleus on staining indicates that, what was taken for a single cell when unstained, was in reality two or even more cells closely attached to one another (*Figure 4B*). Such groups of cells can, when unstained, give the appearance of the 'spider' cells which have been described by DiFrancesco *et al.* (8).

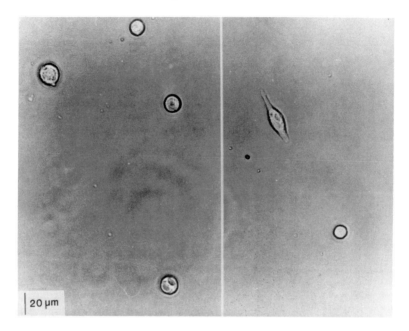

Figure 3. SA node cells in culture. The myocytes all round up but continue spontaneous beating. The cell which is becoming elongated is probably a fibroblast.

4. Electrophysiological study of the isolated sinoatrial node cell

Isolated SA node cells are studied using the patch-clamp technique. In the whole-cell mode, pacemaker activity can be recorded and, under voltage-clamp, whole-cell membrane currents can be registered and analysed. In a cell-attached or cell-detached patch, single channel events can be investigated. The 'macro-patch' has recently been used to study the responses of a small group of I_f channels under cell-detached conditions. This allows direct application of substances to the cytosolic side of the channels (16, 20). Individual I_f channels have such a low conductance (less than 1 pS) that individual channel openings in a conventional patch are difficult to resolve (9).

4.1 Internal solution for the patch pipette

The type of internal solution used in the recording pipette for whole-cell studies is of great importance and may affect the nature and interpretation of

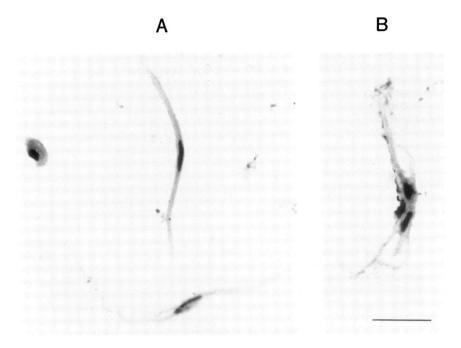

Figure 4. (A) Isolated SA node cells fixed in formalin and stained in haemotoxylin/eosin. Note that each cell has only a single nucleus. (B) This 'spider cell' is shown on fixing and staining to be a group of three cells. Bar represents 30 μm.

the results obtained. The compositions of five internal solutions used by different investigators are compared in *Table 2*.

The solution shown in Column 1 (12, 21) was used for a general survey of all membrane currents; those in Columns 2 and 3 for studies primarily of calcium currents (13, 6) and those in Columns 4 and 5 for investigations of I_f (16, 9). EGTA is included in the pipette solution to prevent Ca^{2+} overload within the cell. At concentrations of 5 mM or 10 mM (6, 13, 16), it will keep the internal Ca^{2+} concentration very low so that any Ca^{2+}-activated currents will be effectively suppressed. Cs^+ in the internal solution is used to block outward K^+ currents (6).

4.2 Whole-cell patch electrode recording from the isolated SA node cell

Spontaneous activity from a rabbit SA node cell is shown in *Figure 2B*. The average maximum diastolic potential (from nine published reports) is 64.5 mV, and the average spontaneous frequency (same nine sources) is 156 beats/min (see reference 22).

Table 2. Composition of internal (pipette) solutions

	Concentration (mM)				
	Solution 1(12)	Solution 2(13)	Solution 3(6)	Solution 4(16)	Solution 5(9)
KCl	140	50		20	
K Aspartate		80		110	130
Aspartic acid			90		
CsOH			110		
CsCl			20		
NaCl					10
$MgSO_4$	3				
$MgCl_2$		8	1	2	
KH_2PO_4		10			
K_2-ATP	3		5	2	
Mg-ATP					2
Na_2-ATP		5			
Trisphosphocreatine	5				
K creatine phosphate			5		
EGTA		10	10	5	1
cyclic AMP		0.1	0.05	0.01	
GTP	0.4			0.1	0.1
Hepes-KOH	11	10		5	10
Hepes-CsOH			5		
pH[a]	7.2	7.4	7.4	7.3	7.2
Pipette resistance (MΩ)[b]	5–9	4–8	3–10	1.5–2	3–5

[a] pH of Solutions 1, 2, and 5 adjusted with KOH, pH of Solution 3 adjusted with CsOH, and Solution 4 adjusted with Tris.

[b] Van Ginneken and Giles (15) used rather higher resistance electrodes (5–20 MΩ) filled with a 150 mM solution of either KCl or 150 mM potassium gluconate to which 10 mM KCl had been added. 10 mM Hepes was added to the solution and the pH adjusted to 7.4 with KOH.

5. Membrane currents

The main time-dependent and time-independent currents of the SA node cell are shown in diagrammatic form in *Figure 5*.

5.1 Time-dependent currents

The chief time-dependent currents can be shown when depolarizing and hyperpolarizing voltage-clamp pulses are given to an SA node cell from a holding potential of about −40 mV (*Figure 6*).

5.1.1 Inward calcium current

In response to the depolarizing voltage-clamp pulses, an inward calcium current (L-type Ca^{2+} current) is first recorded (*Figure 6A*). This is the current underlying the upstroke of the SA node cell action potential in which fast sodium current plays little or no role. It also contributes to the last third of the

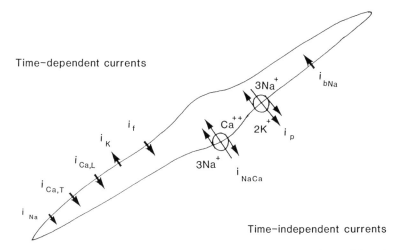

Figure 5. Diagrammatic representation of the membrane currents in an SA node cell. See text for details.

pacemaker depolarization. The increase in L-type Ca^{2+}-current brought about by β-adrenergic agents is a very important factor in accelerating the pacing rate. The presence of another calcium current (transient or 'T-type' calcium current) has been reported in SA node cells (6). Its threshold of activation (-50 mV) is more negative than that of the L-type calcium current, so it could have an important influence on the pacemaker potential. There is, however, too little information available to be certain of the role it plays in SA node pacemaking.

5.1.2 Outward K^+ current

The inward calcium current is succeeded, during depolarizing voltage-clamp pulses (*Figure 6A*), by an outward potassium current (delayed rectifier, I_K). This, after the end of the depolarizing clamp pulse, decays fairly slowly at the holding potential. It is the decay of I_K set against a constant (i.e. time-independent) inward background current ($I_{b,Na}$ see below) which provides one of the depolarizing mechanisms underlying the first part of the pacemaker potential.

5.1.3 Hyperpolarization-activated inward current

The other depolarizing mechanism is provided by the current that is recorded during hyperpolarizing voltage-clamp pulses (*Figure 6B*): the hyperpolarization-activated inward current, I_f. This has what might be thought of as 'ideal' properties for a pacemaker current: it provides more depolarizing drive, the further the membrane is hyperpolarized, and it is very rapidly inactivated at positive potentials (such as those during an action

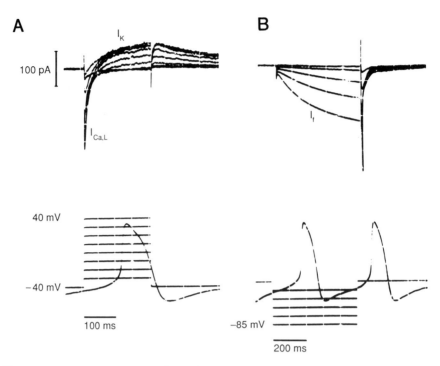

Figure 6. Voltage-clamp pulses (below) and simultaneous membrane current records (above) from an SA node cell. The (unclamped) spontaneous activity of the cell has been superimposed on the voltage-clamp pulses in each case to indicate the level of the pacemaker and action potentials. (A) Depolarizing clamp pulses increasing in 10 mV steps given from a holding potential of −40 mV. The current records (above) show that inward L-type calcium current ($I_{Ca,L}$) is succeeded by outward potassium current, I_K, which decays away quite slowly on return to the holding potential. (B) Hyperpolarizing clamp pulses from a holding potential of −35 mV, again in 10 mV increments, activate the hyperpolarization-activated inward current, I_f.

potential). Nevertheless, its role in SA node pacemaking seems to be contributory rather than essential. Thus, the block of I_f by 2 mM Cs$^+$ does not stop pacemaking, though it will cause considerable slowing of pacemaker rate (21).

5.2 Time-independent currents

The time-independent currents of the SA node cell (shown in *Figure 5*) can be recorded by applying suitable agents to block other membrane currents and by adjusting the conditions and protocol appropriately. Thus the influence of the Na$^+$/K$^+$ pump has been shown in SA node (23), Na$^+$/Ca^{2+} exchange has been detected (Dr. W.-K. Ho, personal communication) and the 'background' Na^+-sensitive inward current, $I_{b,Na}$, has very recently been recorded and analysed (24).

5.3 'Run-down' of membrane currents

Run-down of membrane currents, manifesting itself as a more or less rapid reduction in size of the recorded current, is a problem in single-cell experiments. In single SA node cells, the L-type calcium current is particularly vulnerable to run-down. The run-down occurs in two stages, as is shown in *Figure 7*: a fairly slow decrease in the current over about 15 min, followed by a sudden rapid reduction. The same pattern of run-down has been reported in guinea-pig ventricular cells (25). It has been shown that the run-down of cardiac L-type channels can be reversed by tissue extract containing cytoplasmic proteins (26).

6. Permeabilized patch technique

This variant of the whole-cell patch-clamp technique was introduced by Horn and Marty (27), who used the antibiotic nystatin to make the patch membrane permeable to small cations. Through such a permeabilized patch, 'whole-cell' recording can proceed but loss of larger cell constituents and consequent current run-down is prevented. *Protocol 2* gives instructions for incorporating nystatin into the pipette solution; this has been used for recording from SA node cells (21). More recently, the antibiotic amphotericin has been used for the permeabilization of patches (28, 29). This has the advantage of greater solubility, so that sonication is not necessary.

Protocol 2. Use of nystatin for the permeabilization of membrane patches

Warning: Nystatin is light sensitive, so keep the stock solution and the filling solution pipette and syringe wrapped in black tape. Turn the microscope light off after making a seal.

Method
1. Make a stock solution of 25 mg Nystatin (Sigma) in 0.5 ml dimethyl sulphoxide (DMSO). Keep this solution in the freezer (and in the dark) and make it up fresh every 2 days.
2. Before use, sonicate the stock solution for 6 min.
3. Add 10 μl of the stock solution to 5 ml of filtered pipette solution (0.2 μm filter) in a black-taped test-tube. Shake this and then sonicate it for approximately 12 min before use. Do not re-filter. Make up fresh nystatin-filling solution every 2–3 h.

Protocol 2. *Continued*

4. It can be difficult to get seals in the presence of nystatin, so dip the pipette tip into nystatin-free filling solution for a slow count of seven before back-filling with the nystatin-containing filling solution from a blacked-out syringe. Before filling, discard any nystatin-containing solution which may have been exposed to light (e.g. in the plastic cannula attached to the filling syringe).
5. Permeabilization should occur in about 5 min after seal formation. Monitor permeabilization by applying a 10 mV test pulse. Permeabilization is associated with an increase of both the current amplitude and the capacity transient, which also becomes faster as the access resistance falls. Permeabilization is adequate when the access resistance has fallen to less than 15 MΩ.

Acknowledgements

I should like to thank Dr Jane Denyer for her enthusiastic participation in the development of SA node cell isolation and Dr W.-K. Ho for helpful comments on this manuscript.

References

1. Powell, T. and Twist, V. W. (1976). *Biochem. Biophys. Res. Commun.*, **72**, 327.
2. Powell, T., Terrar, D. A., and Twist, V. W. (1980). *J. Physiol.*, **302**, 131.
3. Taniguchi, J., Kokubun, S., Noma, A., and Irisawa, H. (1981). *Jpn. J. Physiol.*, **31**, 547.
4. Nakayama, T., Kurachi, Y., Noma, A., and Irisawa, H. (1984). *Pflügers Arch.*, **402**, 248.
5. Sakmann, B., Noma, A., and Trautwein, W. (1983). *Nature*, **303**, 250.
6. Hagiwara, N., Irisawa, H., and Kameyama, M. (1988). *J. Physiol.*, **395**, 233.
7. Hagiwara, N. and Irisawa, H. (1989). *J. Physiol.*, **409**, 121.
8. DiFrancesco, D., Ferroni, A., Mazzanti, M. and Tromba, C. (1986). *J. Physiol.*, **377**, 61.
9. DiFrancesco, D., and Tromba, C. (1988). *J. Physiol.*, **405**, 493.
10. Brown, H. F., Campbell, D. L., Clark, R. B. and Denyer, J. C. (1987). *J. Physiol.*, **390**, 60P.
11. Denyer, J. and Brown, H. (1987). *Jpn. J. Physiol.*, **37**, 963.
12. Denyer, J. C. and Brown, H. F. (1990). *J. Physiol.*, **428**, 405.
13. Doerr, T., Denger, R., and Trautwein, W. (1989). *Pflügers Arch.*, **413**, 599.
14. Nathan, R. D. (1986). *Am. J. Physiol.*, **250**, H325.
15. Van Ginneken, A. C. G. and Giles, W. R. (1991). *J. Physiol.*, **434**, 57.
16. Yatani, A. and Brown, A. M. (1990). *Am. J. Physiol.*, **258**, H1947.

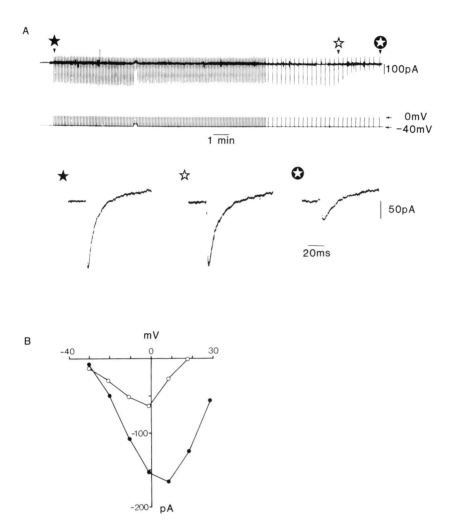

Figure 7. Run-down of L-type calcium current during a whole-cell patch electrode recording from an SA node cell. (A) Top two traces: 200 msec depolarizing voltage-clamp pulses from −40 mV to 0 mV were given once every 5 sec (later once every 15 sec). $I_{Ca,L}$ maintained its amplitude for about 15 min and then ran down rapidly. Lower panel: Current records from times indicated by the symbols on a larger scale. (B) The current–voltage relationship for $I_{Ca,L}$ before (●) and after (○) the rapid phase of run-down. (Records and figure courtesy of Dr Jane Denyer.)

17. Masson-Pévet, M., Bleeker, W. K., Besselsen, E., Treytel, H. J., Jongsma, H. J., and Bouman, L. N. (1982). *J. Mol. Cell. Cardiol.*, **14**, 53.
18. Honjo, H. and Boyett, M. R. (1992). *J. Physiol.*, **452**, 128P.
19. Kohl, P., Kamkin, A. G., Kiseleva, I. S., and Streubel, T. (1992). *Exp. Physiol.*, **77**, 213.
20. DiFrancesco, D. and Tortora, P. (1991). *Nature*, **351**, 145.
21. Denyer, J. C. and Brown, H. F. (1990). *J. Physiol.*, **429**, 401.
22. Irisawa, H., Brown, H. F., and Giles, W. R. (1992). *Physiol. Rev.* (In press.)
23. Noma, A. and Irisawa, H. (1974). *Pflügers Arch.*, **351**, 177.
24. Hagiwara, N., Irisawa, H., Kasanuki, H., and Hosoda, S. (1992). *J. Physiol.*, **448**, 53.
25. Belles, B., Malecot, C. O., Hescheler, J., and Trautwein, W. (1988). *Pflügers Arch.*, **411**, 353.
26. Kameyama, M., Kameyama, A., Nakayama, T., and Kaibara, M. (1988). *Pflügers Arch.*, **412**, 328.
27. Horn, R. and Marty, A. (1988). *J. Gen. Physiol.*, **92**, 145.
28. Rae, J., Cooper, K., Gates, G., and Watsky, M. (1991). *J. Neurosci. Methods*, **37**, 15.
29. Rae, J. L. and Fernandez, J. (1992). *News Physiol. Sci.*, **6**, 273.

4

Electrophysiology of *Xenopus* oocytes: an expression system in molecular neurobiology*

S. P. FRASER, C. MOON, and M. B. A. DJAMGOZ

1. Introduction

Oocytes are unfertilized egg cells which can reach more than 1 mm diameter at later stages of development in lower vertebrates and thereby enable intracellular recording with multiple electrodes to be performed with ease (1). Furthermore, all the biochemical steps, from translation of messenger ribonucleic acid (mRNA) to insertion of protein into the plasma membrane, which normally occur rapidly upon fertilization, may also be induced by injection of 'foreign' mRNAs (2, 3). If the resulting protein is capable of generating an electrophysiological signal, as in the case of neurotransmitter/ hormone receptors, ion channels, and pumps, then electrophysiological recording can be used to assay the induced characteristic. Oocytes of the clawed frog *Xenopus laevis* in particular have been found to be highly efficient in this respect. They have become indispensible in studies aiming at elucidation of the molecular basis of function in proteins, especially those of the nervous system. The aim of this chapter is to give an account of the electrophysiology of *Xenopus* oocytes with an emphasis on practical aspects. An earlier account of this topic ('Functional expression in the *Xenopus* oocyte of mRNAs for receptors and ion channels') was given in 1987 by Barnard and Bilbe in *Neurochemistry: A Practical Approach*.

2. Endogenous characteristics

Before undertaking any electrophysiological experiment involving *Xenopus* oocytes, it is important to be aware of their endogenous characteristics. These characteristics were reviewed extensively in a recent article (4) and only essential details are given here.

* We dedicate this article to Professor Eric Barnard FRS on the occasion of his 65th birthday and for giving us much encouragement with the oocyte technique.

2.1 Morphological properties

Oocytes have a non-uniform structure with two distinct regions: the yellowish 'vegetal' hemisphere and the dark brown 'animal' hemisphere, which are surrounded by several tissue layers (*Figure 1*). The oocyte's membrane gives rise to microvilli which form gap junctions with the macrovilli of the follicle cells. The microvilli, as well as the endoplasmic reticulum, are more abundant in the animal hemisphere, and this is where the oocyte's phosphatidylinositol second-messenger system is concentrated (5). Importantly, the follicle cells also possess receptors and ion channels, and can contribute to the oocyte's overall electrophysiological response.

2.2 Electrophysiological properties

The endogenous electrophysiological characteristics of *Xenopus* oocytes are outlined in *Table 1*.

2.3 Activation of second-messenger systems and G-proteins

As shown in *Table 1*, *Xenopus* oocytes possess both cyclic nucleotide (cAMP and cGMP) as well as phosphatidylinositol (inositol trisphosphate (IP_3), and diacylglycerol (DAG)) second-messenger systems. Of the former,

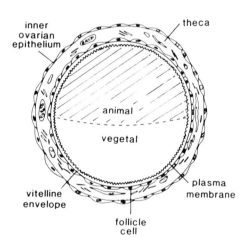

Figure 1. Schematic diagram of morphological features of a Stage V/VI oocyte of *X. laevis* and surrounding tissues. The follicular oocyte is surrounded by a number of cellular and non-cellular layers, as follows (beginning with the innermost): *vitelline membrane envelope*, a non-cellular fibrous layer which remains after collagenase treatment: a monolayer of some 10^4 *follicle cells*—there are gap junctions between the macrovilli of these and the microvilli of the oocyte; *theca*, a fibrous connective tissue layer containing blood vessels, nerve fibres, smooth muscle, and fibroblasts; and an *inner ovarian epithelium*, which is a continuation of the ovary wall. (From reference 45, with permission.)

Table 1. Summary of endogenous membrane characteristics of *Xenopus* oocytes. Characteristics given in parentheses are those for which the evidence is weak

Ion pumps/ exchangers	Ion channels	Equilibrium potentials	Second messengers	Receptors
Na$^+$–K$^+$ ATPase	Fast and slow K$^+$	E_K −100 mv	cAMP	mACh
(Na$^+$–Ca^{2+})	TTX-sensitive Na$^+$	E_{Na} +50 mV	(cGMP)	ß-adrenoceptor
Na$^+$–K$^+$–2Cl$^-$	Voltage-dependent Ca^{2+}	E_{Cl} −22 mV	IP$_3$	Dopamine
Na$^+$–H$^+$	Ca^{2+}-dependent Cl$^-$		(DAG)	(Serotonin)
	Voltage-dependent Cl$^-$			ATP; adenosine
	(Ca^{2+}-dependent K$^+$)			VIP; CRP
				Gonadotropins
				Progesterone
				Angiotensin

TTX, tetrodotoxin; VIP, vasoactive intestinal polypeptide; CRP, corticotropin-releasing peptides.

cAMP has been extensively studied and shown to be associated mostly with a variety of receptors/ion channels present upon follicular cells. Consequently, if the follicular cells become detached from the oocyte by uncoupling of the gap junctions that normally link the two, then cAMP-dependent responses of the oocytes are lost. Uncoupling occurs during prolonged (several days) incubation of the oocytes or as a result of collagenase treatment. On the other hand, the IP$_3$ system is a property of the oocyte's own membrane.

Second messengers can be activated in oocytes by a variety of ways. Levels of cAMP can be elevated by (a) intracellular injection and (b) extracellular application of membrane-permeant analogues (e.g. 8-Bromo-cAMP) or forskolin (an activator of adenylate cyclase). In these situations, it is advisable to include a phosphodiesterase inhibitor (e.g. 3-isobutyl-1-methylxanthine (IBMX)) so as to prevent hydrolysis of cAMP. Direct injection is necessary in order to elevate the level of IP$_3$, since there is as yet no membrane-permeant analogue, but photoactivation of a 'caged' complex may also be used (6). Interestingly, external application of certain divalent cations (e.g. Co^{2+}, Ni^{2+}, Zn^{2+}, Mn^{2+}, and Cr^{2+}) induces Ca^{2+}-dependent Cl$^-$ currents. These presumably raise intracellular Ca^{2+}, possibly by stimulating the IP$_3$ system (7). Zn^{2+} also induces an outward K$^+$ current, which may be indicative of activation of adenylate cyclase (*Figure 2a*). Protein kinase C (normally stimulated by DAG) can be activated by extracellular application of phorbol esters (8).

GTP-binding (G-) proteins of *Xenopus* oocytes have been studied (9). G-proteins can conveniently, but non-specifically, be stimulated by extracellular application of an aluminium fluoride complex (AlF$_4^-$) in Ringer's solution (10, 11). The composition of the 'AlF'-Ringer is shown in *Table 2*. AlF$_4^-$ stimulates G-proteins by mimicking the gamma-phosphate group of GTP (10, 11).

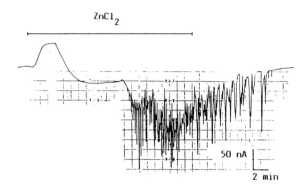

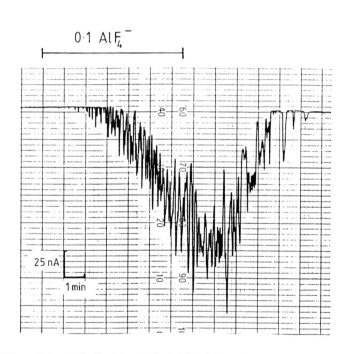

Figure 2. Effect of 1 mM $ZnCl_2$ (a) and '0.1' AlF_4^- (b) on the membrane current in an uninjected oocyte under two-electrode voltage-clamp (holding potential, −60 mV). The durations of applications of $ZnCl_2$ and AlF_4^- are indicated by the horizontal bars above the response traces. (a) $ZnCl_2$ induces two components of membrane current: an initial smooth outward current, probably a cAMP-dependent K^+ current, and subsequent oscillatory Ca^{2+}-dependent Cl^- currents as also seen in (b). (b) After a latency of some 2 min (probably the time taken by AlF_4^- to cross the membrane), oscillatory membrane (Ca^{2+}-dependent Cl^-) currents are induced. The effect can be reversed gradually by washing with normal Ringer's solution. (From reference 4, with permission.)

Table 2. Composition of the stock solution for aluminium fluoride ('AlF') Ringer used for activation of G-proteins in *Xenopus* oocytes

	Concentration (mM)
NaCl	108
KCl	2.5
$CaCl_2$	1.8
Hepes	10
NaF	10
$AlCl_3$	0.1
pH 7.3 (adjusted with 1 M NaOH)	

The stock concentration represents 'I AlF'. Lower concentrations can be prepared and used by diluting this stock solution with normal Ringer's solution.

The effect of AlF_4^- on the membrane current of an oocyte is illustrated in *Figure 2b*. Thus, application of AlF_4^- gradually elicits oscillatory Ca^{2+}-dependent Cl^- currents, which are largely independent of extracellular Ca^{2+} and only partially blocked by pertussis toxin. The currents are similar to those elicited by injection of the second messenger IP_3, or those induced through neurotransmitters coupled to the IP_3 second-messenger pathway. The rise in intracellular concentration of Ca^{2+} occurs probably by release from internal stores, as a result of G- (possibly G_0-) protein activation (9). The 'AlF' technique was used originally for stimulating G-proteins in the characterization of the melatonin receptor expressed in *Xenopus* oocytes (12). Prior stimulation of G-proteins was necessary, since melatonin was coupled to IP_3 in an inhibitory way and, thus, stimulation was a prerequisite for demonstrating receptor expression (see Section 7.4). The same procedure may be useful for studying any expressed receptor which may be inhibitory to a second-messenger system.

3. Strategies for expression studies

3.1 Preparation of mRNA

Numerous protocols for the extraction of RNA are dependent on the origin of the biological material. An RNA extraction procedure for whole-tissue homogenates is the guanidinium thiocyanate method (13). A wide range of poly(A^+) mRNA sizes is then selected and purified on an oligo(dT)– cellulose column (14). However, with the increasing demand for the simplification of the mRNA extraction and purification procedure, there are now many commercial kits which facilitate the process and give relatively good yields. Examples of such kits include:

- for extraction of RNA: RPN 1264 (Amersham); 200345 (Stratagene); 251100 RNAgents (Promega)

Electrophysiology of Xenopus *oocytes*

- for Poly(A$^+$) mRNA isolation: Poly(A)Quick Isolation Kit (Stratagene); Fast Track (Invitrogene); Poly A Tract mRNA Isolation System (Promega); Hybond-mAP (Amersham)

When working with mRNA it is important to minimize the activity of ribonucleases (RNases), which can very easily destroy mRNA activity. To avoid accidental introduction of trace amounts of RNases from potential sources in the laboratory, certain precautions, listed below, should be followed:

(a) *Glassware and plastic-ware.* It is a good idea to set aside items of glassware that are to be used only for preparation of RNA. These items should be treated in an oven (250 °C for 4 h). Disposable, sterile plastic-ware should be used whenever possible.
(b) *Hands of operator.* It is necessary at all times during the preparation of solutions and materials for RNA extraction for the operator to wear disposable latex gloves.
(c) *Solutions.* Solutions should be prepared using RNase-free glassware, autoclaved double-distilled water and the highest quality chemicals. Chemicals should be handled with baked spatulas (250 °C for 4 h). All solutions should be treated by adding 0.1% (v/v) diethyl pyrocarbonate (DEPC) for 12 h prior to use. Traces of this chemical should be removed by autoclaving, since DEPC can react with nucleic acids. DEPC should not be used in Tris solutions because it reacts also with primary amines. Importantly, DEPC is a potential carcinogen and should be handled with care.

Following isolation, it is possible to fractionate the mRNA, most commonly by sucrose density gradients. This gives information on the relative size of the mRNA encoding the protein (and whether subunits encoded by different-sized mRNA are involved). It can increase expression by concentrating the relevant mRNA.

The extracted mRNA is suspended in DEPC-treated sterile distilled H$_2$O at a concentration of 1 µg/µl (or occasionally up to 2 µg/µl) and then frozen in 3 µl aliquots at −70 °C until use. Repeated freezing and thawing of the sample is to be avoided, since this could denature the mRNA.

3.2 Surgery of *Xenopus* and removal of oocytes

It is possible to purchase good-quality, 'sexually-mature' female *Xenopus laevis* frogs from commercial sources in both the UK (Blades Biological) and USA (Nasco Inc.). The frogs are best kept in 15–30 cm of water, preferably in holding tanks that allow the continual circulation of clean, 15–20 °C tap water. The frogs are fed twice weekly on a mixture of 'Xenopus pellets' (a fully balanced diet available from Blades Biological) and chopped liver.

There are two methods of surgical removal of oocytes from *Xenopus* females. The first, and most drastic, involves killing the frog. Owing to the expense of each frog and the inherent variability of different oocyte batches (some batches of oocytes express mRNA more efficiently than others), it is more sensible to re-use the frog following an appropriate recovery time. *Protocol 1* describes the induction of anaesthesia and the subsequent recovery of the frog.

Protocol 1. Anaesthesia of *Xenopus* frogs and removal of oocytes

Materials
- anaesthetic solution: 0.04% (w/v) aqueous solution of ethyl-m-aminobenzoate (also called MS222 or Tricaine)
- Barths' medium (*Table 3*)

Method

1. Anaesthetize a large sexually mature *Xenopus* female in the solution for 15–25 min. Disposable latex gloves should be worn throughout this procedure since the chemical is a possible carcinogen.

2. When fully anaesthetized, place the frog with the ventral side upwards on a damp tissue and cover all but the lower abdomen with tissues soaked in the anaesthetic solution. This action both prevents damage to the frog's mucous membranes from 'drying out' and maintains anaesthesia throughout surgery.

3. With a single cut of a sterile scalpel, make a small (~ 1 cm) incision through the skin above the bladder on one side of the midline. Using sharp surgical scissors, neatly cut through the body muscle into the abdominal cavity. By this action, the ovary should be revealed.

4. Carefully pull several lobes of the ovary through the body muscle, excise, transfer immediately to modified Barths' medium (*Table 3*) and wash several times.

5. Separately suture the body muscle and skin together using a small needle and fine braided silk suture. Use a continuous suture for the body muscle and three or four interrupted inverting sutures for the skin.

6. Following surgery, leave the frog to recover for several hours in a small pool of water, covered by a damp tissue. Support the head above the water to prevent drowning. When fully active, return it to the main tank.

3.3 Preparation of single oocytes

To prepare the oocytes for mRNA injection, it is essential to separate out individual oocytes at maturation stages V–VI (15) from the ovary. There are two different methods.

3.3.1 Manual isolation

Lobes of ovary are placed in Ringer's solution (*Table 3*) and oocytes are mechanically teased out of the lobes using Pasteur pipettes with tips of about 1.5 mm. One pipette is placed over an oocyte and slight suction applied. The other pipette is used to hold the ovary, while the oocyte is gently teased free.

3.3.2 Enzymatic isolation

A lobe of the ovary is placed in Barths' medium (*Table 3*) containing 2 mg/ml collagenase (Sigma, Type 2) at 18–20 °C for 2–3 h. The oocytes can be separated out of the ovary and the surrounding follicular cells by occasional swirling. Enzymatic isolation of the oocytes, unlike manual isolation, dissociates the follicular cells from the oocyte. These cells are coupled via gap junctions to the oocyte; they contain receptors and ion channels, and may also be invovled in the second-messenger pathways. They thus influence the oocytes' endogenous characteristics, and this should be borne in mind when choosing the method of isolation.

Table 3. Composition of *Xenopus* Ringer's solution and incubation (Barths') medium[a]

Chemical	Normal Ringer (mM)	Low-Ca^{2+} Ringer[b] (mM)	Barths' Medium[c]
NaCl	115	115	88 mM
KCl	2.5	2.5	1 mM
CaCl$_2$	1.8	–	0.41 mM
NaHCO$_3$	–	–	2.5 mM
Ca(NO$_3$)$_2$	–	–	0.33 mM
MgSO$_4$	–	–	0.82 mM
Tris–HCl	–	–	7.5 mM
Hepes	5	5	–
Penicillin	–	–	10 µg/ml
Streptomycin	–	–	10 µg/ml
Gentamycin sulphate	–	–	0.1 mg/ml
pH	7.3	7.3	7.4

[a] The solutions are generally made up as stocks (× 10 concentration), autoclaved and stored in a refrigerator. Before use, they are then diluted tenfold, autoclaved, cooled, buffered if required, and antibiotics added.

[b] In this solution, some oocytes may deteriorate rapidly. Stability can be improved by adding 1–2 mM MgCl$_2$. For Ca^{2+}-free solution, 2 mM EGTA is also added.

[c] Additional extras used by some groups include: sodium pyruvate (1 mM), nystatin (50 IU/ml), and fetal calf serum (0.1%).

It is very important at this stage to assess the quality of the oocytes since, as previously stated, there is a variability in successful mRNA expression. It has been found that for the best chance of expression, only the highest quality oocytes should be injected. These oocytes are 'firm' and show a sharp contrast in colour and boundary between the animal (black/dark brown) and vegetal (yellow/creamy) hemispheres. If oocytes show odd pigmentation or are easily broken (this becomes apparent during manual isolation), they should be discarded. It is also useful to impale a few oocytes from a batch and check that they have 'healthy' resting potentials and input impedances. Occasionally, whole batches may have to be discarded, either for lack of the correct stages of oocytes or because they do not pass the 'quality test'.

4. Injection of mRNA

Having isolated a good selection of Stage V–VI oocytes, they should be 'equilibrated' before injection by being left to soak in Barths' medium for 2 h at 20 °C. This enables damaged and dead oocytes to be recognized and discarded. We should reiterate that with any procedures involving the handling of mRNA, or for any equipment that will come into contact with the mRNA, great care should be taken to prevent RNase contamination. This involves the wearing of disposable latex gloves at all times, and the use of individually packaged, sterile plastic tips and pipettes, etc. For glassware, such as the micropipettes, a new box should be opened and its usage restricted solely to experiments involving mRNA.

4.1 Micropipettes

Pipettes for injecting mRNA into the oocytes are pulled from thin-wall glass capillaries (e.g. GC150T-15, Clark Electromedical) in a two-stage pull on a vertical puller. This results in micropipettes which have sharply tapering shanks ideal for injection. It is possible to store the pipettes for a short time before injection in a large sterile disposable Petri dish. Care should be taken to prevent dust, etc. from contaminating the tip.

4.2 Injection apparatus

For a successful injection technique, there are several essential pieces of equipment which can readily be assembled by the operator. These include:

(a) *A micromanipulator and/or a hydraulic microdrive* (e.g. HMD-2M, Clark Electromedical Instruments). These must have high sensitivity to reduce the likelihood of damage to the oocyte during injection.

(b) *A bifocal microscope.* Any good dissecting microscope with variable magnification would be adequate; an inbuilt light source is helpful.

(c) *Light source.* Any microscope lamp will suffice, but drying out of the

Electrophysiology of Xenopus oocytes

oocyte can be a problem, especially for beginners who will take more time over injection. This is solved by using a 'cold' fibre optics light source.

(d) *Injection system.* There are basically two methods of injection:
   i. Manual injection. This covers a wide range of techniques from hand-held to micrometer-controlled syringes.
   ii. Automated injection. This usually involves a microprocessor-controlled fixed or variable volume microdispensor. We have recently tested such an injector (Drummond 3–00–510–X) and found it to be quite satisfactory. This system has the advantage that predetermined volumes of mRNA can be injected repeatedly. There are also commercial devices available for injection of mRNA into oocytes by electroporation (e.g. IBI-Kodak) or 'Baekonization' (Baekon Inc.). However, these have not been used extensively and we are not sure of their quality.

4.3 Injection procedure

An area of the laboratory suitable for carrying out the injection procedure should first be found. It is sensible to restrict the area to mRNA work *only*. If possible, it is wise to contain the necessary equipment within a flow cabinet to reduce the risk of contamination. Before starting the injection, all instruments and the working area should be thoroughly cleaned with 0.1% aqueous DEPC and then wiped down with alcohol. Outlined below is a protocol demonstrating an mRNA injection technique (*Protocol 2*). It must be noted that, depending on equipment used or techniques preferred, the injection procedure varies markedly between groups (16, 17).

Protocol 2. A procedure for the manual injection of mRNA

Wear disposable latex gloves at all times and be wary of possible exposure to ribonucleases.

1. Attach the injection pipette to the injection device via a micro-electrode holder with a pressure port. Bring the tip into focus under the microscope and use a sterile pipette tip to break it carefully to a diameter of 10–15 μm (tips of larger size may damage the oocyte membrane).

2. Thaw a frozen 3 μl aliquot of mRNA (at 1 μg/μl) and pipette it onto the inverted lid of a sterile disposable Petri dish.

3. Draw the mRNA into the injection pipette by applying slight suction. Using the microscope, watch for blockage of the tip. This can be prevented by alternate 'sucking' and 'blowing' actions.

4. Transfer 4–5 oocytes using a Pasteur pipette to the inside edge of a sterile 50 mm Petri dish lid. Draw off the excess incubation medium. With the tip

of the pipette, group the oocytes together and orient them such that the vegetal poles are approachable. Leave enough medium, however, to prevent drying out.

5. Bring the injection micropipette into view and touch the tip to the middle of the exposed surface of the oocyte's vegetal pole.
6. With a gentle tap and turn of the micromanipulator, the oocyte membrane will indent and then return to its original position, indicating impalement.
7. Inject 40–70 nl of mRNA into each oocyte; watch for slight swelling of the oocyte (this confirms mRNA has been injected).
8. Remove the micropipette and repeat the procedure on the remaining oocytes. It is important that the animal pole (the dark pigmented hemisphere) is not injected since this part contains the easily damaged nucleus.

Although it is important to carry out the steps of *Protocol 2* as quickly as possible to prevent the oocytes from drying out, mRNA injection should be performed slowly to prevent damage and excessive stretching.

4.4 Maintenance of oocytes

Following injection, groups of four to five oocytes should be kept in small (50 mm) Petri dishes in modifed Barths' medium (*Table 3*). Although not essential, it is recommended that they are placed in an ordinary incubator at 20–21 °C. Both the incubation medium and the Petri dish should be changed daily, using the flow cabinet to ensure sterility, and oocytes showing signs of infection should be discarded.

It is essential that uninjected and water-injected oocytes are maintained and regularly tested as controls for the mRNA preparations.

Electrophysiological investigation of the oocytes usually begins 1–5 days after mRNA injection. The period of the initial incubation before screening and characterization is determined by trial and error, and according to the nature of the expression.

5. Electrophysiologial recordings

Membrane recordings from mRNA-injected oocytes enable the characterization and study of the functional properties of expressed voltage- and/or ligand-gated ion channels. Two main electrophysiological recording techniques are used:

- two-electrode voltage clamp
- patch-clamp

5.1 Two-electrode voltage-clamp

This is the most widely used means of recording from oocytes, whereby one electrode records the membrane potential while the other injects current. Many large companies market the necessary equipment (e.g. Axoclamp 2-A, Biologic CA-100/VF-180, and Dagan 8500/8800). A limiting factor with this technique is that the circuit must be able to produce large ($\sim$ 10 µA) currents necessary to voltage-clamp the oocyte.

The large size of the oocyte facilitates impalement with two electrodes, but unfortunately the inherent large capacitance ($\sim$ 300 nF) limits the clamping speed. To combat this, the electrode resistances are kept low at 1–3 MΩ and pipette capacitance is reduced by maintaining the bath level close to the surface of the oocyte. Inserting a grounded metal shield between the electrodes decreases any capacitative coupling.

The intracellular microelectrodes are pulled from small-diameter glass capillary tubing containing an internal filament to facilitate filling (e.g. GC100F-10, Clark Electromedical). Any good-quality puller will suffice, so long as the electrodes have tip resistances of 1–3 MΩ when filled with 2.5 M KCl.

For most ligand-gated receptor studies, voltage and current recordings can be followed on a high-quality chart recorder. However, if voltage-gated channels are studied (when membrane events are much faster in the millisecond range), a computer-controlled data acquisition package (e.g. CED 1401, Cambridge Electronic Design; pCLAMP, Axon Instruments) is essential for digital storage of records and data analysis.

Upon impalement of the oocytes with the first microelectrode, a variable resting potential of between -30 and -80 mV is usually recorded. Insertion of the second microelectrode results in the resting potential usually falling to around -20 mV, but when left for 10–15 min there is a gradual increase of both resting potential (from -30 to -50 mV) and membrane resistance (from 0.5 to 4 MΩ). The cell can then be voltage-clamped and tested for receptor expression. Most agonist applications are performed at a holding potential of -60 mV. Steady-state 'current–voltage' ($I-V$) relationships are obtained by stepping the potential to various levels before, during, and after the application of the agonist.

5.2 Patch-clamp

Patch-clamp recording has been used mostly in structure–function studies of expressed receptors or ion channels, whereby information about single channel kinetics may be acquired. Since this technique requires a clean membrane surface, the oocyte needs to be prepared specially. The steps used for obtaining a 'naked' oocyte suitable for patch-clamping are illustrated in *Figure 3* and outlined in *Protocol 3*.

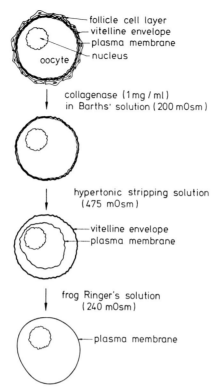

Figure 3. Schematic representation of the steps used for the mechanical skinning of oocytes (normal or mRNA-injected), resulting in exposure of the 'naked' plasma membrane for patch-clamp recording. The initial collagenase treatment leaves only the vitelline membrane, which is removed finally by applying the hypertonic stripping solution. (From reference 18, with permission.)

Protocol 3. Preparation of *Xenopus* oocytes for patch-clamp (from reference 18, see also *Figure 3*)

Materials

- Barths' medium (*Table 3*)
- stripping solution (mM): K aspartate, 200; KCl, 20; $MgCl_2$, 1; EGTA, 10; Hepes, 10; pH 7.4

Method

1. Two days after injection of mRNA, incubate oocytes in 1 mg/ml collagenase (Sigma Type I) in Barths' medium for 1 h (19 °C), and then wash in Barths' medium.

2. Remove the follicle layer mechanically with a pair of fine forceps.

Protocol 3. *Continued*

3. Immediately before patch-clamping, place the oocyte for 5–10 min (20–22 °C) into 'stripping' solution. The osmolarity of this solution is 475 mOsm and hypertonic to Barths' medium (200 mOsm).
4. Remove the vitelline membrane completely once it becomes detached from the plasma membrane of the oocyte, using a pair of forceps (*Figure 3*). Note that without the follicular cells and the vitelline membrane, the oocytes are extremely fragile and should be handled with care.

Protocol 3 describes the mechanical removal of the oocytes' vitelline membrane; an enzymatic method involving protease treatment has also been described (18).

Patch-clamp recordings can be performed in a number of modes (whole-cell, cell-attached, inside-out, outside-out patches, etc.). Details of obtaining such recordings and the patch pipettes and their filling solutions are outside the scope of this chapter, but can readily be found in several other publications (19, see also Chapters 1, 2 and 8). Many manufacturers who produce voltage-clamp apparatus also make equipment for patch-clamping (e.g. Biologic RK300, Dagan 8900, and List EPC7/9). Similarly, several software packages are available for control of patch-clamp recordings and data analysis.

5.3 Perfusion system

For voltage- or patch-clamping, the oocyte is placed in a small perfusion chamber ($\sim$ 0.25 ml) and continually perfused with frog Ringer's solution (*Table 3*) at a rate of approximately 3 ml/min at room temperature (19–20 °C). The excess solution is drawn off by vacuum pump to a waste flask, via a micropipette attached to plastic tubing. The perfusion system allows application of up to seven different solutions using a centrally controlled multi-valve network. The dead-time for the perfusion system is around 6 sec. An alternative to the perfusion of agonists is 'instant' application by using the 'U-tube' technique (20) or pressure injection of the drug. In the latter situation, a micropipette with a broken tip is placed in the proximity of the oocyte. A controlled pressure pulse is then applied to release a fixed quantity of agonist in a limited area. This technique may be used in cases of receptors showing suspected desensitization to bath-applied agonists. The bath is grounded with a silver/silver chloride wire or, to avoid possible junction potentials, an agar bridge made up with Ringer's solution may be used.

6. Complementary screening techniques

In addition to electrophysiological testing, there are several alternative techniques for the screening of proteins expressed in oocytes. Some of these rely on binding of specific ligands or toxins, e.g. α-bungarotoxin upon nicotinic cholinergic receptors and ouabain upon the α-subunit of Na^+/K^+ ATPase (21). When enzymes are expressed, then biochemical assay of the relevant activity in oocyte homogenates may be used (22). A screening technique of particular relevance to phosphatidylinositol-linked receptors expressed in *Xenopus* oocytes involves fluorometric measurement of intracellularly released Ca^{2+}. *Protocol 4* describes an assay for detecting changes in intracellular Ca^{2+} levels using the photoprotein aequorin (23). This luminometric assay is well suited for structural/functional analysis of receptors, agonist/antagonist studies and, especially, initial and rapid screening of expression libraries.

Protocol 4. Luminometric assay for detection of phosphatidylinositol-linked receptor expression in *Xenopus* oocytes

Materials
- aequorin: 1 mg/ml in 1 mM EDTA, stored in aliquots at −85 °C
- low calcium Ringer's solution
- Barths' medium

Method
1. Dry down the mRNA sample.
2. Resuspend the mRNA in an equal volume of aequorin.
3. Rinse the injection needles with 1 mM EDTA.
4. Inject some oocytes with 40–60 nl RNA/aequorin and control oocytes with 40–60 nl low-calcium Ringer's solution (see *Table 3*).
5. Incubate the oocytes for 24 h at 18 °C in Barths' medium (*Table 3*).
6. Determine the oocyte's response to the ligand by pipetting single oocytes into 400 μl Ringer's solution in 12 × 55 mm disposable polystyrene tubes and placing them in the luminometer or photon counter.
7. Determine the oocyte's response to the addition of agonist in 50 μl Ringer's solution and compare this with the effect of 50 μl Ringer's solution alone.

Other Ca^{2+}-sensitive dyes/indicators include fura-2 and flou-3 (24, 25). With the use of a fluorescence microscope or fluorimeter, the internal Ca^{2+} concentration of the cell can be monitored. Although the use of Ca^{2+}-sensitive dyes is limited by their restriction to receptors linked to the phosphatidylinositol second-messenger pathway, other indicators (sensitive to Na^+, H^+, or cAMP) may similarly be used. A potential advantage of this approach is that it can cut down on the screening time of mRNA-injected oocytes, especially when a large number of fractions are involved.

7. Recent expression studies

Utilizing recombinant DNA technology, which has enabled the cloning and sequencing of complete proteins, it is now possible to isolate from cDNA clones the mRNAs encoding for specific subunits. This has resulted in numerous studies characterizing the function of specific domains or subunits of proteins expressed in *Xenopus* oocytes. It is not possible within the scope of this chapter to give an extensive account of these studies. We have instead summarized representative examples of the different strategies adopted in this approach, as follows.

7.1 Expression and characterization of specific subunits

By injection of cRNAs (mRNA synthesized from cloned DNA) encoding for one or more of the subunits of a specific receptor/channel, the functional role of the various subunits can be investigated. From studies of the γ-aminobutyric acid (GABA) receptor, $GABA_A$, which is composed of up to four different polypeptide subunits (α, β, γ, δ), some of which have several variants, it has been found that expression of either α or β subunits separately in *Xenopus* oocytes resulted in the formation of a functional channel with characteristics similar to those of the native receptor (26). These results suggest that both α and β subunits can substitute for each other to form a functional homo-oligomeric channel complex. However, receptors formed by the combination of α and β subunits desensitized more rapidly, showed greater outward rectification and displayed smaller single-channel conductances than those assembled from α and γ subunits (27). $GABA_A$ receptors are also known to be the target for a variety of chemicals, including barbiturates, benzodiazepines, steroid anaesthetics, and alcohol. Benzodiazepine (e.g. diazepam) enhancement of GABA responses was obtained with either the γ2S or γ2L subunit, whilst barbiturate (e.g. pentobarbital) action did not require any specific subunit. Ethanol enhancement of GABA action was seen only if the expressed receptor contained the γ2L subunit (in addition to α and β subunits). Furthermore, the ethanol sensitivity of the $GABA_A$ receptor complex depended not only on the presence of the γ2L subunit but also on the phosphorylation state of this subunit (28).

In contrast to the $GABA_A$ receptor, work with muscle nicotinic acetylcholine (nACh) receptors indicated that, of the four subunits (α, β, γ, δ), most were necessary to form a functional channel (29, 30). Only when γ or δ subunits were separately omitted did the receptor still show some, albeit weak, activity.

7.2 Co-translation of mRNAs from different sources

The vertebrate muscle nACh receptor is a protein complex consisting of four separate subunits with the stoichiometry $αβ_2γδ$. Expression of mouse–*Torpedo* 'hybrid' receptors in *Xenopus* oocytes, by injecting a mixture of the mRNAs encoding each subunit, enabled a detailed characterization of subunit contribution to the overall receptor properties. By comparing the channel conductance values of different combinations, the γ and δ subunits were found to influence single channel conductance (31, 32). Similarly, mouse α and δ subunits lengthened channel open time, whilst the β subunit shortened it. Voltage-dependency was found to reside in the β and δ subunits.

Cat muscle nACh receptors were found to desensitize more slowly than analogous receptors from *Torpedo* electric organ when expressed separately in *Xenopus* oocytes. In order to investigate the molecular basis of this difference, cat-*Torpedo* hybrid nACh receptors were formed by injecting oocytes with mixed mRNAs. Hybrid nACh receptors formed by co-injection of cat muscle mRNA with the *Torpedo* β- or δ-subunit mRNA desensitized as slowly as before. However, when the cat mRNA was co-injected with *Torpedo* γ-subunit mRNA, densensitization of the hybrid nAChR was much faster. It could thus be concluded that the difference in the rates of densensitization of cat and *Torpedo* nACh receptors is determined mainly by the respective γ subunits (33).

Studies of these kinds are thus very important in analysing the functional architecture of ion channel and receptor proteins. This aproach also enables determination of potential differences in receptor characteristics occurring during development, e.g. the nACh receptor changes character during development with an alteration of the γ subunit in fetal muscle to ε in adult muscle (34).

7.3 Translation of mRNAs following site-directed mutagenesis

Upon cloning the voltage-sensitive sodium channel, it was discovered that it contains four homologous repeats, each consisting of six putative transmembrane segments. It was proposed, from the largely positive residue composition of one of these segments (S4), that this area is involved in the depolarization-induced activation of the channel. A similar situation probably exists in voltage-dependent potassium and calcium channels. In order to test this hypothesis, point mutations were introduced into the region of S4, thereby either neutralizing the positive residues or replacing them with negative ones (35, 36). The resulting channels were expressed and characterized in *Xenopus* oocytes. The net effect of reducing the positive charges in S4 was indeed a reduction in the voltage-dependence of activation. Further experiments with deletion mutants at the linkage between repeats III and IV caused a reduction in the rate of inactivation of the channel, thereby implying that this region is important for channel closure. Thus, studies involving deletions and mutations in channels expressed in *Xenopus* oocytes can greatly facilitate the functional analysis of specific regions of channel proteins. A review of structure–function studies of voltage-gated ion channels was published recently (37).

Similar mutation studies, using the α-bungarotoxin-sensitive, homo-oligomeric, neuronal nicotinic α7 ACh receptor, investigated the effects of varying the highly conserved leucine 247 residue of the MII region (38). All the mutants formed functional receptors with identical Hill coefficients and sensitivity to α-bungarotoxin, indicating that the overall structure of the receptor was unaltered. However, mutations to amino acid residues which were smaller and/or more hydrophobic than the wild type resulted in a reduction in the rate of desensitization and current rectification, as well as an increase in the affinity for ACh. In addition, leucine-to-threonine mutations induced an additional conductance state at low ACh concentrations. These effects were considered possibly due to mutants causing a high-affinity desensitization state (normally closed in the wild-type) to become open. The results indicate the importance of the leucine 247 residue in the inherent functioning of the ion channel. In addition they suggest that, because of its highly conserved nature in all nicotinic, glycine, and $GABA_A$ receptors, the residue may have a related role in these receptors.

7.4 Expression of a receptor that inhibits a second messenger

Melatonin is known to be a neurohormone important in controlling regulation of both circadian and photoperiodic functions in many vertebrate species, with recent research suggesting also potential links with depression, schizophrenia, and disorders such as cancer. The cellular mechanism of action of the melatonin receptor has, however, not been well characterized. Most research has centred on the action of melatonin in the pars tuberalis (PT) of the pituitary. In sheep PT cells, signal transduction is achieved via two G-proteins, one pertussis toxin insensitive, the other pertussis toxin sensitive, linked to inhibition of cAMP accumulation.

Injection of poly(A^+) mRNA, isolated from ovine PT cells, into *Xenopus* oocytes resulted in the expression of melatonin receptors (12). Characterization of this receptor was achieved using AlF_4^- to stimulate G-proteins (see Section

2.3). Application of AlF_4^- to an oocyte resulted in an oscillatory current with a reversal potential of -20 mV (*Figure 2b*), close to the Cl^- equilibrium potential in *Xenopus* oocytes. In PT mRNA-injected oocytes clamped at -60 mV, melatonin alone showed no effect on the membrane current. However, when applied during AlF_4^- stimulation, melatonin reversibly abolished the oscillatory current. Uninjected oocytes stimulated by AlF_4^- showed little or no reduction in the oscillatory current to melatonin application. The effect of melatonin was dose-dependent in the 50 μM to 1 mM range and showed little desensitization, even at high concentrations.

These results indicate the expression of a functional melatonin receptor, possibly coupled via G-proteins to a signal transduction system with the characteristics of an IP_3/calcium second-messenger system. Interestingly, there was lack of fidelity in the G-protein coupling of the expressed melatonin receptors, since in oocytes the receptors coupled to the IP_3 system, whilst in native PT cells melatonin receptors couple to cAMP.

7.5 Expression of neurotransmitter receptors from invertebrates

There is now considerable evidence to suggest that major pharmacological differences exist in receptors to the same endogenous ligand in vertebrates and invertebrates. It would, therefore, be of considerable interest and importance to determine the molecular bases of these dissimilar characteristics. Determination of the molecular structures resulting in the differences between the receptors will be particularly relevant to the pharmaceutical and agrochemical industries when considering the design of novel and more specific chemical control agents, e.g. pesticides and insecticides.

7.5.1 Glutamate receptors

Recent studies have attempted to focus on these inherent differences. In an initial characterization (39), amino acid receptors were expressed in *Xenopus* oocytes using muscle mRNA from the locust (*Shistocerca gregaria*). Membrane currents were detected to applications of GABA, glutamate, quisqualate, and ibotenate. The latter two agonists are known to activate, respectively, the D-type and H-type of postjunctional glutamate receptors at insect excitatory synapses. Observed differences of these amino acid receptors from ones from vertebrate sources included severe desensitization of the glutamate receptors. Often, oocytes became insensitive to the agonists following a single application, even prolonged washing being ineffective in returning the response.

7.5.2 GABA receptors

Although the locust GABA receptor gated Cl^- and was inhibited by picrotoxin, it was insensitive to both baclofen and bicuculline, indicating that it is distinct from the vertebrate $GABA_A$ and $GABA_B$ receptors (39).

In order to characterize the differences at the molecular level, much effort has been levelled at cloning the genes coding for invertebrate receptors. Recently, a subunit of the GABA receptor from the mollusc *Lymnaea stagnalis* has been sequenced (40). This polypeptide had about 50% sequence homology with the vertebrate $GABA_A$ receptor β-subunit and, when expressed in *Xenopus* oocytes, it formed functional GABA-activated Cl^- channels which were blocked by bicuculline. Interestingly, the molluscan subunit was able to assemble with the bovine $GABA_A$ α1-subunit to form functional receptors. The evidence suggests that the molluscan receptor is of the $GABA_A$ type and probably exists *in vivo* as a hetero-oligomeric complex.

7.5.3 *Drosophila* receptors

A glutamate receptor subunit from *Drosophila melanogaster* has also been cloned (41). The subunit, designated DGluR-II, showed between 37 and 38% homology to the rat glutamate receptor subunits, GluRl, 4, and 5. Electrophysiological recordings of the subunits expressed in *Xenopus* oocytes revealed currents up to 500 nA to extremely high concentrations of L-glutamate (100 mM) and L-aspartate (100 mM) and a reversal potential of −10 mV. Quisqualate, AMPA, and kainate were relatively ineffective, as were the glutamate antagonists, argiotoxin, 6-cyano-7-nitroquinoxaline-2,3-dione, 6,7-dinitroquinoxaline-2,3-dione, and D-APV. The agonist sensitivity for the subunit was, however, low with an EC_{50} of 35 mM. This may be due, however, to the lack of complementary subunits usually present and necessary for complete functional expression.

7.5.4 Nicotinic receptors

Work with the invertebrate nACh receptor has similarly discovered a receptor with considerable differences from that of the vertebrate. Cloning of a putative receptor from the locust, *S. gregaria*, resulted in the isolation of cDNA (termed 'αLl'), which encodes a single α-subunit type of polypeptide sequence (42). The predicted size of the mature polypeptide from this αLl clone corresponds well to a 65K protein purified by α-bungarotoxin (α-BuTX) affinity from the locust, *Locusta migratoria* (43).

Expression of the locust αL1 cDNA in *Xenopus* oocytes resulted in receptors which, when activated by micromolar concentrations of nicotine, gated a non-specific cation channel. The observed response was reversibly blocked by mecamylamine (100 µM), δ-tubocurarine (1 µM), and bicuculline (100 µM). In addition, α-BuTX (0.1 µM) completely blocked the response to nicotine (10 µM). Not only do these results concur with a previous report of an nACh receptor, composed of a single subunit, which was expressed in *Xenopus* oocytes from neural tissue of *L. migratoria* (44), but the observed pharmacology of the expressed putative αLl nACh receptor also agrees with *in vivo* data, and the sensitivity to bicuculline and α-BuTX indicates differing characteristics from the vertebrate neuronal nACh receptor.

References

1. Hagiwara, S. and Jaffe, L. A. (1979). *Ann. Rev. Biophys. Bioeng.*, **8**, 385.
2. Gurdon, J. B., Lane, C. D., Woodward, H. R., and Marbaix, G. (1971). *Nature*, **233**, 177.
3. Sumikawa, K., Houghton, M., Emtage, J. S., Richards, B. M., and Barnard, E. A. (1981). *Nature*, **292**, 862.
4. Fraser, S. P. and Djamgoz, M. B. A. (1992). In *Current Aspects of the Neurosciences*, Vol. 4 (ed. N. N. Osborne), pp. 267–315. Macmillan, Basingstoke.
5. Berridge, M. J. (1988). *J. Physiol.*, **403**, 589.
6. Parker, I. and Miledi, R. (1989). *J. Neurosci.*, **9**, 4068.
7. Miledi, R., Parker, I., and Woodward, R. M. (1989). *J. Physiol.*, **417**, 173.
8. Lupu-Meiri, M., Shapira, H., and Oron, Y. (1989). *Eur. J. Physiol.*, **413**, 498.
9. Moriarty, T. M., Pedrell, E., Carty, D. J., Omri, E., Landau, E. M., and Iyengar, R. (1990). *Nature*, **343**, 79.
10. Sternweiss, P. C. and Gilman, A. G. (1982). *Proc. Natl. Acad. Sci. USA*, **79**, 4888.
11. Bigay, J., Deterre, P., Pfister, C., and Chabre, M. (1987). *EMBO J.*, **6**, 2907.
12. Fraser, S. P., Barrett, P., Djamgoz, M. B. A., and Morgan, P. J. (1991). *Neurosci. Lett.*, **124**, 242.
13. Chomoszynski, P. and Sacchi, N. (1987). *Anal. Biochem.*, **162**, 156.
14. Aviv, H. and Leder, P. (1972). *Proc. Natl. Acad. Sci. USA*, **69**, 1408.
15. Dumont, J. N. (1972). *J. Morphol.*, **136**, 153.
16. Coleman, A. (1984). In *Transcription and Translation: A Practical Approach* (ed. B. D. Hames and S. J. Higgins) pp. 271–302. IRL Press, Oxford.
17. Boyle, M. B. and Kaczmarek, L. K. (1991). *Methods Neurosci.*, **4**, 157.
18. Methfessel, C., Witzemann, V., Takahashi, T., Mishina, M., Numa, S., and Sakmann, B. (1986). *Eur. J. Physiol.*, **407**, 577.
19. Sakmann, B. and Neher, E., (eds) (1983). *Single Channel Recording*. Plenum Press, New York.
20. Suzuki, S., Tachibana, M., and Kaneko, A. (1990). *J. Physiol.*, **421**, 645.
21. Hara, Y., Ohtsubo, M., Kojima, T., Noguchi, S., Nkao, M., and Kawamura, M. (1989). *Biochem. Biophys. Res. Commun.*, **163**, 102.
22. Wall, D. A. and Patel, S. (1989). *J. Membr. Biol.*, **107**, 189.
23. Giladi, E. and Spindel, E. R. (1991). *BioTechniques*, **10**, 744.
24. Brooker, G., Seki, T., Croll, D., and Wahlestedt, C. (1990). *Proc. Natl. Acad. Sci. USA*, **87**, 2813.
25. Lechleiter, J., Girard, S., Peralta, E., and Clapham, D. (1991). *Science*, **252**, 123.
26. Blair, L. A. C., Levitan, E. S., Marshall, J., Dionne, V. E., and Barnard, E. A. (1988). *Science*, **242**, 577.
27. Seeburg, P. H., Wisden, W., Verdoorn, T. A., Pritchett, D. B., Werner, P., Herb, A., Luddens, H., Springel, R., and Sakmann, B. (1990). *Cold Spring Harbor Symp. Quant. Biol.*, **LV**, 29.
28. Wafford, K. A., Burnett, D. M., Leidenheimer, N. J., Burt, D. R., Wang, J. B., Kofuji, P., Dunwiddle, T. V., Harris, R. A., and Sikela, J. M. (1991). *Neuron*, **7**, 27.
29. Mishina, M., Kurosaki, T., Tobimatsu, T., Morimoto, Y., Noda, M., Yamamoto,

Y., Terao, M., Lindstrom, J., Takahashi, T., Kuno, M., and Numa, S. (1984). *Nature*, **307**, 604.
30. White, M. M., Mayne, K. M., Lester, H. A., and Davidson, N. (1985). *Proc. Natl. Acad. Sci. USA*, **82**, 4852.
31. Sakmann, B., Methfessel, C., Mishina, M., Takahashi, T., Takai, T., Kurasaki, M., Fukuda, K., and Numa, S. (1985). *Nature*, **318**, 538.
32. Yu, L., Leonard, R. J., Davidson, N., and Lester, H. A. (1991). *Mol. Brain Res.*, **10**, 203.
33. Sumikawa, K. and Miledi, R. (1989). *Proc. Natl. Acad. Sci. USA*, **86**, 367.
34. Mishina, M., Takai, T., Imoto, K., Noda, M., Takahashi, T., Numa, S., Methfessel, C., and Sakmann, B. (1986). *Nature*, **321**, 406.
35. Stühmer, W., Conti, F., Suzuki, H., Wang, X. and Noda, M., Yahagi, N., Kubo, H., and Numa, S. (1989). *Nature*, **339**, 597.
36. Papazian, D. M., Timpe, L. C., Jan. Y. N., and Jan, L. J. (1991). *Nature*, **349**, 305.
37. Stühmer, W. (1991). *Annu. Rev. Biophys. Biophys. Chem.*, **20**, 65.
38. Revah, F., Bertrand, D., Galzi, J.-L., Devillers-Thiery, A., Mulle, C., Hussy, N., Bertrand, S., Ballivet, M., and Changeux, J.-P. (1991). *Nature*, **353**, 846.
39. Fraser, S. P., Djamgoz, M. B. A., Usherwood, P. N. R., O'Brien, J., Darlison, M. G., and Barnard, E. A. (1990). *Mol. Brain Res.*, **8**, 331.
40. Harvey, R. J., Vreugdenhil, E., Zaman, S. H., Bhandal, N. S., Usherwood, P. R. N., Barnard, E. A., and Darlison, M. G. (1991). *EMBO J.*, **10**, 3239.
41. Schuster, C. M., Ultsch, A., Schloss, P., Cox, J. A., Schmitt, B. and Betz, H. (1991). *Science*, **254**, 112.
42. Barnard, E. A., Darlison, M. G., Marshall, J., and Sattelle, D. B. (1989). In *Ion Transport* (ed. D. Keeling and C. Benham) pp. 159–181. Academic Press, London.
43. Breer, H., Kleene, R., and Hinz, G. (1985). *J. Neurosci.*, **5**, 3386.
44. Breer, H., and Benke, D. (1986). *Mol. Brain Res.*, **1**, 111.
45. Snutch, T. P. (1988). *Trends Neurosci.*, **11**, 250.

PART II
Synaptic transmission

5

Pharmacological analysis of synaptic transmission in brain slices

GRAEME HENDERSON

1. Introduction

For the study of synaptic transmission in the mammalian central nervous system (CNS) the *in vitro* brain slice preparation presents a significant number of advantages over *in vivo* techniques.

(a) Foremost amongst these is the relative ease with which recordings can be obtained using extracellular, intracellular and whole-cell patch-clamp techniques.

(b) In certain slice preparations (e.g. the hippocampus), one can see under a low-power dissection microscope the presynaptic fibre tracts, the areas of synaptic contact and the cell body layer; thus, the stimulating and recording electrodes can be positioned at the desired site under visual control.

(c) Since the brain slice is maintained in an artificial bathing solution, the ionic composition of the extracellular environment can be manipulated.

(d) With intracellular and whole-cell patch-clamp recording, the ionic composition of the intracellular milieu can also be controlled.

(e) Drugs can be applied to the synapse under study in known concentrations by addition to the bathing solution.

It was originally thought that patch-clamp recordings could only be made from neurones maintained in tissue culture because a clean membrane surface onto which the patch pipettes could be sealed was essential. More recently it has been realized that patch-clamp recordings can be obtained from neurones in a brain slice and this has re-invigorated brain slice experiments. Furthermore, with whole-cell voltage-clamp recording from neurones contained in brain slices, quantal analysis of synaptic transmission (see Chapter 6), can now be performed on central synapses at a degree of resolution previously only possible on peripheral neurones or CNS neurones maintained in culture.

Synaptic transmission at many sites in the CNS has been examined in brain slices. In this chapter are described commonly used techniques for the study of synaptic transmission in slices (see also Chapter 8). The term *postsynaptic response* will be used to describe postsynaptic events resulting from presynaptic stimulation and which could be recorded in the postsynaptic neurone, either as a postsynaptic potential (i.e. in current-clamp recording) or as a postsynaptic current (i.e. in voltage-clamp recording).

2. Electrical recording from brain slices

2.1 Preparation of brain slices

Over the 30 years since Henry McIlwain first pioneered the use of brain slice preparations for electrophysiological experiments on CNS neurones, a myriad of diverse techniques have been developed for the cutting and maintenance of brain slices. In the early 1980s much effort and discussion was focussed on determining the best techniques to use (see, for example, the final chapter in *Brain Slices* (1)). No single approach has been adopted and people tend to use the techniques which work for them. Outlined in *Protocol 1* is the procedure for cutting brain slices, which I and numerous other colleagues have developed and found to be successful with such diverse brain regions as the locus coeruleus, substantia nigra, ventral tegmental area, nucleus accumbens, striatum, and hippocampus. There are certainly different approaches taken in other laboratories (e.g. see reference 2).

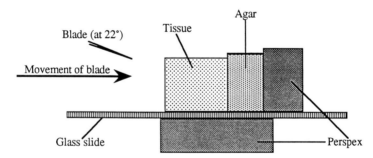

Figure 1. Brain slice sectioning. The tissue block is mounted on a glass slide against an agar wall. The Vibratome blade is advanced (indicated by the arrow) at an angle of 22 ° to the horizontal plane.

Protocol 1. Brain slice preparation

Equipment

- a Vibratome (Oxford Instruments) or equivalent for slicing unfixed brain tissue

- a modified chuck for the Vibratome comprising a glass microscope slide onto which are glued two small Perspex blocks: one on the bottom of the slide for clamping the chuck in the Vibratome, the other to provide a wall against which to mount the agar and tissue blocks (see *Figure 1*)

Method

1. Using stout scissors and/or bone cutters, remove the skin and upper skull without damaging the brain.
2. Using the tip of a scalpel blade, cut the dura overlying the brain along the midline and retract it on either side. With the scalpel blade, make a firm cut through the spinal cord at its junction with the medulla.
3. Using a small spatula gently lift the brain from the cranial cavity.
4. Place the brain on a piece of filter paper moistened with bathing solution.[a]
5. Keeping the brain tissue moist with bathing solution, prepare a block of tissue containing the region from which recordings are to be made. Accomplish this with firm cuts of a sharp razor blade.
6. Mount the tissue block on the chuck of the Vibratome with one side resting against the agar block (*Figure 1*). Attach the tissue and agar block to the chuck using cyanoacrylate adhesive (Superglue).
7. Place the mounted tissue in the cutting chamber of the Vibratome such that, after cutting through the brain tissue, the blade will enter the agar block. The chamber should contain bathing solution at 4 °C, continually bubbled with O_2/CO_2.
8. Cut serial slices of the desired thickness (80–750 μm) by advancing the cutting blade of the Vibratome at a slow speed and with a large, lateral blade movement ('vibration'). The recommended blade angle is 22 ° to the horizontal. It is essential that the blade passes smoothly through the tissue—like a hot knife through butter—and that the blade does not push against and compress the tissue block. Ragged tissue, blood vessels, and blood clots all hinder the cutting of good slices.
9. Transfer the slices carefully from the Vibratome to the recording chamber or to a storage chamber. A fine paintbrush and a broad, bent spatula are useful for this purpose.
10. Store slices before transfer to the recording chamber by placing them in a beaker containing bathing solution at room temperature and constantly bubbling the solution with O_2/CO_2. Take care not to agitate the slices as the solution is bubbled.

[a] Step 4 and all of the following procedures are done using bathing solution at 4 °C in order to cool the tissue, thus reducing oxygen demand and rendering brain tissue, such as the brain stem and spinal cord, sufficiently firm to enable cutting of slices. For robust tissue, such as hippocampus and cortex, cooling may not be necessary.

2.2 Recording chamber

As with the preparation of brain slices, a plethora of techniques have been devised for maintaining and recording from brain slices *in vitro*. We have found that fully submerged brain slices of 80–500 μm thickness can be maintained at temperatures up to 37 °C and remain viable for periods in excess of 12 h. The criteria for viability are:

- the magnitude of the resting membrane potential of the neurones
- the stability of the membrane potential
- the threshold, height, and shape of spontaneous or evoked action potentials
- the nature of the postsynaptic response evoked by stimulation of presynaptic inputs

Successful intracellular and whole-cell patch-clamp recordings can be obtained fron neurones contained in slices resting on a net (i.e. both upper and lower surfaces in contact with the bathing solution), as in the Scottish Chamber (described in the final chapter of reference 1), or in slices laid on a glass plate (i.e. only the upper surface in contact with the bathing solution). This latter chamber is made simply by gluing four small pieces of Perspex to a glass microscope slide, such that they form the walls of a well of approximately 0.5 ml in volume into which the slice can be placed and covered with superfusing solution. With both methods it is important that movement of the tissue be prevented by placing on the upper surface of the slice a grid or nylon mesh, which is held in place by small weights or a loop of platinum wire attached to a micromanipulator. Recording and stimulating electrodes can be positioned on the slice through the grid or nylon mesh.

2.3 Bathing solution

It is most common in brain slice experiments to use a bicarbonate-buffered bathing solution equilibrated with 95% O_2/5% CO_2, although phosphate- or Hepes-buffered solutions can also be used. Although such solutions are commonly referred to as artificial cerebrospinal fluid, their chemical composition bears little resemblance to physiological cerebrospinal fluid; rather they are solutions in which brain slices can be maintained *in vitro*. We have used successfully the solution described in *Table 1*, and have found that a constant flow rate through the chamber of between 1.5 and 2 ml/min provides sufficient oxygen to maintain viable slices. Pulsation-free perfusion can be obtained using 'Newton pumps', i.e. gravity! The inflow is driven by a constant pressure head adapted from a classical Marriotte bottle and the outflow is syphoned off.

Table 1. Bathing solution

	Concentration (mM)
NaCl	126
KCl	2.5
NaH_2PO_4	1.24
$MgCl_2$	1.3
$CaCl_2$	2.4
$NaHCO_3$	26
Glucose	10

[a] The solution is saturated with O_2 and the pH maintained at 7.2 by bubbling with 95% O_2/5% CO_2 before use.

[b] When the experiment is to be performed at temperatures above room temperature, the bathing solution is warmed just prior to entering the recording chamber. To avoid the gas mixture coming out of solution and thus unwanted bubble formation in the recording chamber as the solution is warmed, saturate the solution with the gas mixture at the temperature it will be used during the experiment, not at room temperature.

[c] Use of $MgCl_2$ rather than $MgSO_4$ permits the subsequent addition of ions such as Ba^{2+} whose sulphate salts are insoluble.

2.4 Methods of evoking and recording postsynaptic responses

2.4.1 Stimulation of presynaptic fibres

The technique we use for stimulating presynaptic inputs in brain slices is described in detail in *Protocol 2*. When stimulating presynaptic elements it is most desirable to be able to stimulate identifiable fibre tracts which innervate the postsynaptic neurones under study. This can be achieved, for example, in slice preparations of the hippocampus, olfactory cortex, and spinal cord, where the afferent fibre tracts can be identified under low-power magnification and a focal-stimulating electrode placed directly upon the appropriate tract. However, in other brain regions (e.g. locus coeruleus, ventral tegmental area, and substantia nigra the afferent pathways can no longer be identified after the slice has been cut. In these preparations, the stimulating electrodes can be placed close (within 1 mm) to the cell from which the recording is being made. The current field created by the stimulus passed between the two poles of the stimulating electrode is adjusted such that it is sufficient to excite the synaptic inputs to that cell. A description of the multicomponent postsynaptic response evoked by this method in neurones contained in the locus coeruleus is given in Section 3.2.2.

Protocol 2. Stimulation of synaptic inputs

Equipment

- a stimulator with graded DC analogue output
- a stimulus isolation unit (constant voltage or constant current) with the capability for polarity reversal

Pharmacological analysis of synaptic transmission in brain slices

Protocol 2. *Continued*

- a stimulating electrode. Use either a bipolar, etch-sharpened tungsten or nichrome wire electrode (pole separation 0.3 mm), insulated with a varnish coating to within 0.1 mm of the tips. Or, for thin brain slices visualized under Nomarski optics, use a saline-filled large-tipped microelectrode which provides a more convenient focal-stimulating electrode. Current is passed between this electrode and a second, Ag/AgCl wire electrode placed in the bathing solution

Method

1. To avoid dislodging the recording electrode from the cell, place the stimulating electrode in the desired place on the surface of the slice prior to impaling the neurone.
2. Advance the tips or tip of the stimulating electrode into the slice (< 100 μm should be sufficient).
3. After a successful impalement has been obtained, apply single, isolated, square wave current pulses across the stimulating electrode at a low frequency (once every 10–30 sec).[a] Current pulses are passed through the isolation unit to isolate them from ground and thus reduce the stimulus artefact.
4. Increase the amplitude of the current pulse until a suitable postsynaptic response is observed.[b,c,d]
5. Avoid using such large currents that electrolysis occurs (the electrode tips become covered in small bubbles of gas), as this may damage the tissue and will increase the electrical resistance between the electrodes.

---

[a] Synaptic responses may run down with time due to transmitter depletion. For this reason the stimulation rate should be kept as low as possible.

[b] In most instances, increasing the strength of the stimulus current increases the number of presynaptic fibres activated and thus increases the amplitude of the postsynaptic response. The amplitude of the postsynaptic response may also be increased by delivering brief trains of stimuli (e.g. five pulses at 10–100 Hz). These may be required to evoke slow postsynaptic responses or to increase small responses. Alternatively, the postsynaptic response may be increased in amplitude by moving the membrane potential away from the equilibrium potential of the postsynaptic response. This is done in current-clamp by injecting direct current into the cell and in voltage-clamp by altering the holding potential.

[c] Synaptic responses to not reverse in polarity if the stimulus polarity is reversed, but their stimulus artefacts do!

[d] When the stimulating electrodes are placed close to the cell from which the recording is being made, the current applied across the stimulating electrodes may directly depolarize the cell soma and evoke an action potential in the cell before a synaptic input is evoked. Avoid this by changing the polarity of the stimulus or by moving the stimulating electrode further away from the cell.

In certain experiments, it would be advantageous to be able to stimulate a single presynaptic axon, and thus evoke in the postsynaptic cell a postsynaptic response originating from a single presynaptic input. This could be achieved by impaling two neurones, one of which synapses on to the other, with intracellular electrodes. Thus, selective excitation of the first neurone, by applying sufficient depolarizing current through the intracellular electrode to evoke an action potential in this cell, will elicit in the second neurone a postsynaptic response originating from the release of neurotransmitter only from the stimulated neurone. Obviously this is only appropriate for slices in which two cell groups are synaptically coupled, and would not be possible with long projecting neurones. Single-fibre postsynaptic responses have been studied by stimulation of individual hippocampal CA3 pyramidal neurones recording from one of their target neurones in the CA1 pyramidal region. However, this is a difficult technique to use successfully. An alternative approach, used with high-resolution patch-clamp recording, is to use such a low intensity of stimulus that, in repeated trials, it either evokes or fails to evoke the minimum response measurable (see Section 3.4.1). This stimulus is then assumed to be stimulating a single presynaptic fibre.

2.4.2 Extracellular recording

Depending upon the aims of the experiment, it may be sufficient to study the postsynaptic response occurring in a population of cells rather than the response in a single cell. This may be done using DC extracellular recording techniques. The advantage of this approach is in the relative ease and long duration of recording. In tissues with a defined architecture (e.g. the hippocampus), where the area of synaptic contact of the presynaptic terminal with the dendritic tree of the postsynaptic neurone can be identified, then extracellular DC recording of excitatory postsynaptic potentials can be achieved by placing a glass recording pipette filled with bathing solution or 4 M NaCl (tip resistance 5–15 MΩ) either in the region of the synapses or the cell bodies. When the recording electrode is placed in the dendritic layer in the region of the synapses, a single presynaptic stimulus gives rise to a complex waveform (*Figure 2*) comprising:

(a) the stimulus artefact

(b) the presynaptic volley (the action potential evoked in the presynaptic fibres)

(c) the population excitatory postsynaptic response (negative-going in extracellular DC recordings)

(d) a positive-going population spike, if the evoked excitatory post-synaptic potential is sufficiently large to reach threshold for action potential initiation in the postsynaptic neurones.

If the recording electrode is placed in the cell-body layer, then the waveform recorded in response to the same presynaptic stimulus is altered.

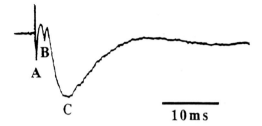

Figure 2. Extracellular recording of the synaptic response recorded from CA1 hippocampal pyramidal neurones. The trace illustrates a subthreshold synaptic response obtained with the recording electrode placed in the dendritic layer. A, stimulus artefact; B, presynaptic volley; C, excitatory postsynaptic response, normally 1–2 mV in amplitude. (Reproduced and modified from Klapstein and Colmers (1992) *Br. J. Pharmacol.*, **105**, 470, with permission.)

The excitatory postsynaptic potential now appears as a positive-going potential and the population spike as a negative-going potential.

In the hippocampus it is also possible to record population postsynaptic potentials using a variant of the grease-gap technique (3). In a transverse hippocampal section, the alveus can be partially separated from the remainder of the slice by a scalpel cut, running from the subiculum to the mid-region of the CAI pyramidal cell layer. A wick electrode is then placed on the cut end of the alveus with a grease barrier separating it from the remainder of the slice. The population postsynaptic response evoked by stimulation of the Schaeffer collateral–commissural pathway can then be recorded as the potential difference between this electrode and another placed on the other side of the barrier. For this type of recording, the slice is not maintained in a recording chamber but lies on an inclined platform, and bathing solution is perfused onto its upper surface.

2.4.3 Single-electrode current- and voltage-clamp

Using a single, intracellular microelectrode, the postsynaptic response can be recorded in individual neurones either as a postsynaptic potential by current-clamp recording or as a postsynaptic current by single-electrode voltage-clamp recording (SEVC). Because of the small size of most CNS neurones and the difficulty in visualizing individual cell bodies in slices > 150 μm thick, the technique of two-electrode voltage-clamp in which the cell is impaled with two independent microelectrodes is somewhat impractical. SEVC is best done with as low a resistance microelectrode as possible, < 50 MΩ is recommended, although successful recordings can be obtained with higher resistance electrodes. In SEVC the intracellular microelectrode functions both as the recording and current-passing electrode, usually with a duty cycle of 70% voltage recording and 30% current passing. Sampling frequencies up to 10 kHz can be used, although care should be taken to monitor the head stage voltage in order to ensure adequate settling of the clamp. The response characteristics of the SEVC can be improved by keeping the depth of fluid above the slice to a minimum and coating the shank of the electrode down to near its tip with an insulating material such as Sylgard. For a detailed description of SEVC see reference 4.

The mode of recording chosen depends to a large extent on the objectives of the experiment. Current-clamp represents the more physiological situation. In current-clamp, but not voltage-clamp, the problem of non-linear summation of the postsynaptic response arises (5). As the strength of stimulus applied to the presynaptic input is increased, then increasing amounts of transmitter are released. However, in current-clamp the amount of potential change evoked in the postsynaptic neurone does not increase linearly with the amount of transmitter; as the potential changes in response to the transmitter, it moves towards the equilibrium potential for the response and thus reduces the driving force for subsequent potential change as the amount of transmitter is increased. This does not apply in voltage-clamp, because with this technique the membrane voltage remains constant and the driving force is constant, since it is the current flow in response to the transmitter that is measured.

2.4.4 Whole-cell patch-clamp recording

The applicability of patch-clamp recording to the study of CNS neurones and technical information on how to obtain such recordings is discussed in detail in Chapter 8. Here we are only concerned with the use of whole-cell patch-clamp in the study of synaptic transmission. Synaptic currents can be recorded using either the thick (individual cells not visualized) or thin (individual cell somata visualized under Nomarski optics) brain slice technique. We have found that whole-cell recording from thick (> 150 μm) brain slices is simpler to perform and recordings are obtained more frequently than with conventional sharp intracellular recording techniques. In comparison to single-electrode voltage-clamp, whole-cell voltage-clamp provides a significantly improved signal-to-noise ratio. It is therefore better suited for the study of small amplitude events such as spontaneous miniature synaptic currents (see Section 3.4.1). Indeed, in small electrically compact neurones, such as cerebellar granule cells, it is even possible when using whole-cell recording to observe not only spontaneous miniature excitatory synaptic currents and evoked synaptic currents, but also single channel openings in response to synaptically released transmitter (6).

3. Analysis of postsynaptic potentials

3.1 Blockade of synaptic transmission

To ensure that the response recorded in the postsynaptic neurone, as a result of applying a current pulse across the stimulating electrodes, is indeed synaptic in origin and not some form of electrical artefact, it is necessary to demonstrate that no response is evoked under conditions in which transmitter release is abolished. A non-selective block of transmission at all synapses in the CNS can be produced by preventing calcium entry into the presynaptic nerve terminal. This can be done by removing Ca^{2+} from the extracellular solution, but Ca^{2+} removal may also result in destabilization of the postsynaptic cell membrane and a resultant membrane depolarization. A more convenient procedure is to use inorganic ions which compete for and block the voltage-dependent calcium channels in the presynaptic nerve terminals. Raising the extracellular $[Mg^{2+}]$ to 20 mM or addition of Cd^{2+} (300 μM) or La^{3+} (100 μM) without altering the extracellular $[Ca^{2+}]$ completely blocks neurotransmitter release. Agents such as ω Conus toxin GVIA, dihydropyridine antagonists such as nifedipine and nimodipine, and Funnel Web Spider toxins block specific types of voltage-sensitive calcium channel and, although they can be used to examine the calcium channel type(s) involved in excitation–secretion coupling, they may not necessarily produce complete block of transmitter release.

3.2 Pharmacological dissection of multi-component postsynaptic potentials

Stimulation of afferent inputs may give rise to multi-component postsynaptic potentials in two ways:

- release of a single neurotransmitter on to more than one receptor type present on the postsynaptic neurone
- release of more than one transmitter either from the same or different nerve terminals

3.2.1 A single transmitter acting on more than one receptor
i. *Excitatory amino acids*

The major component of excitatory synaptic transmission in the mammalian CNS appears to be mediated by excitatory amino acids. Whilst the exact nature of the transmitter is still in doubt, the most likely candidates are L-glutamate and L-aspartate. The postsynaptic actions of excitatory amino acids are mediated by a number of receptor types classified primarily according to the most selective exogenous agonist at each receptor type, (i.e. N-methyl-D-aspartic acid, (NMDA), α-amino-3-hydroxy-5-methyl-4-isoxazolepropionic acid (AMPA), kainate, L-2-amino-4-phosphonobutyric acid (AP4)). These receptors are in fact receptor–ion channel complexes (ionotropic receptors), the NMDA receptor-channel being permeable to Na^+, K^+, and Ca^{2+}, whereas the AMPA and kainate receptor-channels are permeable to Na^+ and K^+. A distinct, metabotropic receptor for excitatory amino acids has also been described. The development of selective pharmacological antagonists for some of these receptors (D-2-amino-5-phosphonopentanoic acid (APV) and

3-((±)-2-carboxypiperazin-4-yl)propyl-1-phosphonic acid (CPP) for NMDA receptors, 6-nitro-7-sulphamoyl-benzo(f) quinoxaline-2,3-dione (NBQX) for AMPA receptors, 6-cyano-7-nitroquinoxaline-2,3-dione (CNQX) for AMPA and kainate receptors) and the characteristic enhancement of NMDA responses in solutions containing zero Mg^{2+} has allowed their contribution to excitatory synaptic transmission to be assessed. In the presence of extracellular Mg^{2+}, responses mediated through the NMDA receptor are depressed over the potential range of -40 to -100 mV due to Mg^{2+} block of the NMDA channel and, thus, NMDA responses are described as exhibiting a region of negative slope conductance.

Excitatory amino acid-mediated neurotransmission has been extensively studied at a number of synapses in the CNS. Because of its involvement in the phenomenon of long-term potentiation, a putative model of memory, excitatory amino acid-mediated synaptic transmission has been most extensively studied in the hippocampus. Stimulation of the hippocampal Schaeffer collateral–commissural pathway evokes both excitatory postsynaptic potentials (EPSPs) and inhibitory postsynaptic potentials (IPSPs) in CA1 pyramidal neurones (*Figure 3*) (7). With low intensities or low frequencies of stimulation the EPSP is abolished by CNQX, indicating that it

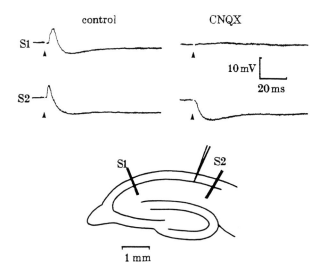

Figure 3. Excitatory and inhibitory postsynaptic potentials recorded in CA1 hippocampal pyramidal neurones. Stimulation of the Schaeffer collateral–commisural pathway at a site distant from the recording electrode (S1) evoked both excitatory and inhibitory postsynaptic potentials which were all abolished by application of CNQX. In contrast, stimulation at a site closer to the recording electrode (S2) directly activated both the Schaeffer collateral–commisural input and the inhibitory interneurones and thus only the excitatory postsynaptic potential is abolished by CNQX, revealing the directly activated biphasic inhibitory postsynaptic potentials. (Reproduced from reference 7, with permission.)

is mediated by AMPA or kainate receptors. However, increasing the stimulus current intensity results in an increase in EPSP amplitude and duration associated with the appearance of a CNQX-resistant component of the EPSP. This longer lasting, CNQX-resistant component is augmented in Mg^{2+}-free solutions, increases in amplitude as the membrane is depolarized from -90 to -50 mV and is abolished by APV. It is thus mediated by NMDA receptors. An NMDA component to synaptic transmission can also be observed following stimuli of high frequency (100 Hz) and low intensity. These two components of the EPSP result from the same transmitter, presumably glutamate, being released from the presynaptic nerve terminals to activate two pharmacologically distinct types of receptor.

ii. Inhibitory amino acids

The major inhibitory transmitter in the CNS, γ-aminobutyric acid (GABA), activates both $GABA_A$ and $GABA_B$ receptors. The $GABA_A$ receptor is an ionotropic receptor permeable to Cl^-. In CNS neurones, the intracellular Cl^- concentration is normally low, and the equilibrium potential for Cl^- (E_{Cl^-}) lies around -70 mV. However, if the intracellular recording pipette is filled with 3 M KCl (KCl is frequently chosen as the electrode-filling solution because the ionic mobilities of K^+ and Cl^- are similar in aqueous solutions), then the gradual leak of Cl^- from the pipette into the cell will raise the intracellular Cl^- concentration and move E_{Cl^-} to more positive potentials. Thus, an initially hyperpolarizing postsynaptic response mediated by an increased Cl^- permeability may be converted into a depolarizing response as E_{Cl^-} changes during the experiment. When studying GABA-mediated postsynaptic responses, it may be more appropriate to use potassium acetate- or potassium methylsulphate-filled electrodes and thus maintain E_{Cl^-} constant. This also applies to another inhibitory amino acid neurotransmitter, glycine, which activates a Cl^- conductance. Postsynaptic responses mediated through the $GABA_B$ receptor results from an increased conductance to K^+, although $GABA_B$ receptors have also been observed to be negatively coupled to Ca^{2+} channels.

In the hippocampus, focal stimulation, close to the CA1 pyramidal neurone from which the recording is being made, can activate directly inhibitory GABA-releasing non-pyramidal interneurones as well as the excitatory input described above (*Figure 3*) (7). The resulting IPSP is comprised of a fast $GABA_A$ component and a slower $GABA_B$ component. This can be demonstrated by selectively blocking each component with appropriate pharmacological antagonists, such as bicuculline and picrotoxin for $GABA_A$, and phaclofen and 2-hydroxysaclofen for $GABA_B$. If the stimulating electrode is now placed further away (> 1 mm) along the Schaeffer collateral–commissural pathway, the biphasic GABA-mediated IPSP can still be evoked, but now both components can be blocked by application of the excitatory amino acid antagonist, CNQX. This would indicate that these IPSPs are being activated polysynaptically by feedforward and/or recurrent

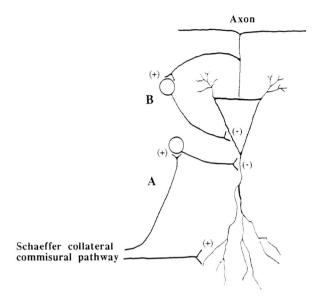

Figure 4. Schematic representation of feedforward (A) and feedback (B) inhibitory synaptic pathways in the CA1 pyramidal cell region of the hippocampus. (+) indicates an excitatory synapse at which the neurotransmitter is most likely glutamate; (−) indicates an inhibitory synapse at which GABA is the neurotransmitter.

feedback pathways in which excitatory amino acids are also involved (*Figure 4*).

3.2.2 Release of multiple neurotransmitters

Focal stimulation of the afferent inputs to the nucleus locus coeruleus evokes a biphasic postsynaptic response, which in current-clamp recording consists of a fast depolarizing potential followed by a slower hyperpolarizing potential (*Figure 5*) (8). Three presynaptically released neurotransmitters are involved in this complex postsynaptic response. It would be incorrect to describe the fast depolarizing component as an EPSP, since it is in fact the sum of a fast EPSP and a fast IPSP. The fast EPSP is mediated by excitatory amino acid receptors and, in the presence of extracellular Mg^{2+}, is approximately 20% NMDA receptor-mediated and 80% non-NMDA receptor-mediated. The fast IPSP results from activation of postsynaptic $GABA_A$ receptors and can be blocked by the $GABA_A$ antagonist, bicuculline. The reversal potential (E_{rev}) of the $GABA_A$-mediated postsynaptic potential is around −50 mV, when chloride-filled recording electrodes are used, but around −70 mV when acetate-filled electrodes are used. Thus, at a membrane potential of −60 mV, the $GABA_A$ IPSP will be depolarizing and summate with (chloride-filled recording electrode), or be hyperpolarizing and attenuate (acetate-filled recording electrode), the bicuculline-insensitive fast EPSP. Similar results

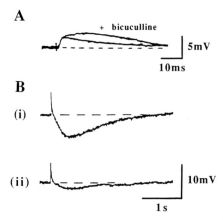

Figure 5. Excitatory and inhibitory postsynaptic potentials recorded in locus coeruleus neurones. In (A) the recording electrode was filled with potassium acetate. In the presence of bicuculline (10 μM) the amplitude of the depolarizing synaptic potential was increased because the fast, hyperpolarizing $GABA_A$ receptor-mediated inhibitory postsynaptic component was removed. In (B) (i) and (ii) the recordings were made on a slower time scale and the slow inhibitory postsynaptic potential is now apparent. In (ii) the slow inhibitory postsynaptic potential was markedly reduced in the presence of the α_2-adrenoceptor antagonist, yohimbine (100 nM). (Reproduced and modified from references 8 and 9, with permission.)

have been observed at other CNS synapses, such as those on to striatal principal neurones and dopaminergic neurones of the ventral tegmental area.

In locus coeruleus neurones, the slow IPSP results from opening of K^+ channels following activation of α_2-adrenoceptors, not $GABA_B$ receptors (9). The slow IPSP is blocked by the α_2-adrenoceptor antagonist, idazoxan, and mimicked by the α_2-adrenoceptor agonists, noradrenaline, clonidine, and adrenaline. E_{rev} for both the slow IPSP and the response to exogenous α_2-adrenoceptor agonists is close to the equilibrium potential for K^+. The identity of the neurotransmitter responsible for the slow IPSP is still unknown. In order to evoke the synaptic input to locus coeruleus neurones, the stimulating electrodes have to be placed close to the cell and presynaptic nerve terminals stimulated. It could be that adrenaline-containing fibres originating in the nucleus paragigantocellularis have been stimulated or that neighbouring locus coeruleus neurones have been excited to release noradrenaline from axon collaterals, somata, or dendrites.

3.2.3 Receptor identification

In the above descriptions of the pharmacological analysis of synaptic transmission, antagonists were used to identify the postsynaptic receptors involved. Such studies rely on the specificity of the antagonists chosen and, in brain slice experiments, some degree of confidence regarding that specificity can be obtained from knowing the concentration of antagonist applied in the fluid bathing the cell. More extensive pharmacological characterization of a receptor requires the measurement of the affinity of agonist and antagonist drugs for the receptor. Antagonist affinity can be determined from the rightward shift of the agonist concentration–response curve produced by increasing concentrations of the antagonist. It is more difficult to measure the affinity of agonists. This requires the use of irreversible antagonists to remove any spare receptor reserve. Such studies can be performed on brain slices (10) and, although somewhat laborious, are required for rigorous receptor identification.

3.3 Ionic mechanism

The interaction of a neurotransmitter with its postsynaptic receptors gives rise either to a change in the membrane permeability to one or more ions or to activation of intracellular enzymes. Here we are only concerned with the changes in membrane ionic conductance. Postsynaptic responses can result from a transmitter-evoked increase in the permeabilitiy of a single ion species (e.g. K^+ or Cl^-), an increase in the permeability through a single channel type of more than one ion species (e.g. Na^+ and K^+ or Na^+, K^+, and Ca^{2+}) or a decrease in the permeability of an ion species (Na^+ or K^+). This last usually is associated with slower, modulatory postsynaptic responses. To study the ionic basis of a postsynaptic response, it is necessary to isolate it from any other contaminating postsynaptic responses. This can be done by addition to the bathing medium of appropriate pharmacological antagonists to block out contaminating responses or by selective placement of the stimulating electrode to ensure that only a single type of presynaptic input is being activated.

3.3.1 Membrane conductance measurement

During a postsynaptic response, the postsynaptic membrane becomes more or less permeable to certain ions depending upon the action of the neurotransmitter involved. With longer duration postsynaptic responses, the change in membrane conductance can be measured under current-clamp recording by repeatedly applying small (< 10 mV), constant amplitude, short duration (around 150 msec) hyperpolarizing current pulses and measuring the change in membrane potential they evoke before, during, and after the postsynaptic response. From Ohm's Law, if the electrotonic potential increases during a postsynaptic potential, then the membrane conductance must have been decreased by the action of the neurotransmitter, i.e. channels in the membrane have been closed. The converse is the case for postsynaptic potentials resulting from an increase in membrane conductance.

3.3.2 Reversal potentials

Characterization of the change in ionic conductance underlying a postsynaptic response can be made by determining the reversal potential (E_{rev}) of that response. To do this the postsynaptic response is evoked repeatedly, whilst the membrane potential of the postsynaptic neurone is altered by applying direct current through the recording electrode in current-clamp (this is effectively the same as changing the holding potential in voltage-clamp). As the membrane potential is moved towards E_{rev}, the driving force for current flow is decreased and the amplitude of the postsynaptic response will decrease. At E_{rev} no current flow will occur. Thereafter, further change in the membrane potential will reverse the direction of the postsynaptic response. The reversal potential of the postsynaptic response will be a function of the equilibrium potential(s) of the ion (or ions), whose permeability has been altered by the action of the neurotransmitter. The equilibrium potential for each ion is dependent upon its concentration across the cell membrane and is the potential at which the electrical gradient completely opposes the chemical gradient. With whole-cell patch-clamp recording, the ion concentrations inside the cell are known, since the cell will be dialysed with the pipette-filling solution and the reversal potentials can be calculated from the Nernst equation. With sharp electrode intracellular recording, these can only be estimated or determined by experiment, (e.g. the after-hyperpolarization following an action potential is due to increased K^+ conductance and will reverse at the K^+ equilibrium potential). In intact mammalian CNS neurones bathed in a standard bathing solution, the equilibrium potentials for the following ions are normally in the region of: Na^+, +50 mV; K^+, −90 mV; Ca^{2+}, > +100 mV; Cl^-, −70 mV. Obviously, if a transmitter-activated channel is permeable to more than one ion species, then the reversal potential of that response will also be dependent upon the relative permeabilities of the different ions through the channel. Erroneous values for E_{rev} may be obtained if the postsynaptic response is evoked in the postsynaptic neurone at sites on the dendritic tree distant from the cell soma. Depending on the electrotonic length of the dendritic tree and the distance from the soma of the synaptic inputs, then injection of current into the soma through the recording electrode to change the membrane potential may evoke a lesser change in potential at distant sites of initiation of the postsynaptic response.

Further information regarding the ionic conductances underlying a postsynaptic response can be obtained by examining the effect on the amplitude of the postsynaptic response and its reversal potential of (a) changing the extracellular concentration of an ion or (b) replacing individual ions thought to be involved in the response with other impermeant ions. Thus, if an inhibitory postsynaptic response appears to be due to opening of a K^+ channel, then intracellular K^+ can be replaced with Cs^+ in whole-cell patch-clamp recordings. With sharp electrode intracellular recording, even injection of small amounts of Cs^+ into a cell from the recording electrode can be sufficient to block K^+ channels. When studying postsynaptic responses, it is more difficult to replace extracellular sodium with, for example, N-methyl-D-glucamine, since this will also block the action potential in the presynaptic fibres. However, to circumvent this difficulty, one could examine the effect of sodium substitution on the response evoked in the postsynaptic neurone by exogenously applied neurotransmitter, rather than on the synaptic response itself. Choline can also be used to substitute for Na^+, but it stimulates muscarinic receptors and, if it is likely that the cell possesses muscarinic receptors, should be used in conjunction with a muscarinic antagonist such as atropine.

3.4 Sites of action of drugs modifying synaptic transmission

3.4.1 Quantal analysis

Quantal analysis provides a powerful tool with which to study how changes in synaptic transmission are produced by drugs or adaptive processes. It is generally accepted, but not definitively proven, that neurotransmitter release occurs in multi-molecular packets of transmitter molecules of constant size (quanta) and that a postsynaptic response results from the synchronous release of a number of quanta. Katz and colleagues, working on the neuromuscular junction (see reference 11 for review), demonstrated that, when the probability of transmitter release from the motoneurone was lowered by reducing extra-cellular $[Ca^{2+}]$ and raising extracellular $[Mg^{2+}]$, the amplitude of the evoked postsynaptic response occurred in steps equal in amplitude to that of the spontaneous miniature end plate potentials. These are the postsynaptic responses to spontaneously released single quanta of transmitter. At the neuromuscular junction, the amplitudes of the evoked responses and the number of failures can be shown to conform to a Poisson distribution. A drug acting presynaptically to reduce transmitter release would be expected to reduce the number of quanta released (*quantal content*) by a given stimulus, but not to affect the amplitude of the postsynaptic response evoked by a quantum of transmitter (*quantal size*). Conversely, a drug acting postsynaptically to antagonize the action of the neurotransmitter would reduce the quantal size but not affect the quantal content. Similarly, with adaptive changes such as long-term potentiation or depression, it should be possible to discriminate between a pre- or postsynaptic locus underlying the adaptive change by quantal analysis (but see below). It should be noted, however, that there are exceptions to the above, since agents or processes which act presynaptically to deplete neurotransmitter content would reduce quantal size and thus might appear to be acting postsynaptically.

In CNS neurones, quantal analysis is complicated by the fact that a single neurone can receive multiple synaptic inputs from numerous presynaptic neurones and that these may be located at various sites, distant from the soma. It is, therefore, necessary to study electrically compact neurones and to stimulate only a single presynaptic input. With whole-cell patch-clamp recording, a higher signal-to-noise ratio is obtained and thus very small amplitude synaptic currents can be recorded. The stimulus intensity can be adjusted to be just sufficiently strong to evoke the smallest synaptic current measurable. The amplitude of the evoked synaptic current will vary and occasionally will be observed as a failure. The presynaptic stimulus is then assumed to be exciting only a single presynaptic fibre. When a measure of the spontaneous miniature synaptic currents is to be made, tetrodotoxin must be present to abolish synaptic events resulting from the firing of spontaneous action potentials in the presynaptic terminals.

Recently, the assumption that, at synapses in the CNS, the size of the quantal event is determined by the amount of transmitter released has been challenged (12). From analysis of the amplitudes of spontaneous miniature inhibitory synaptic currents and evoked inhibitory synaptic currents in cells of the hippocampal dentate gyrus, it has been proposed that at this synapse the quantal nature of synaptic events arises not from the amount of transmitter released, but from the limited availability of receptors on the subsynaptic membrane opposed to each release site. Thus, at each synaptic contact, the presynaptic terminal may indeed release a single quantum of transmitter, but the amount of transmitter released is in excess of the number of available postsynaptic receptors. The amplitude of the postsynaptic current will thus depend on the number of receptors, not on the amount of transmitter released. In addition, other investigators (13) have questioned whether excitatory transmitter release in the hippocampal CA1 region can be described by either a Poisson or simple binomial distribution. If these results prove to be correct and universally applicable to synaptic transmission throughout the CNS, then a much more rigorous quantal analysis will be required to distinguish between pre- and postsynaptic modulation of synaptic transmission. A more detailed discussion of quantal analysis is given in Chapter 6.

3.4.2 Presynaptic inhibition

The amplitude of the postsynaptic response may be used as a sensitive, but indirect, measure of the amount of neurotransmitter released from the presynaptic nerve terminals. This experimental approach provides an estimate of the amount of transmitter released by a single presynaptic stimulus, which can rarely be obtained by other more direct methods of measuring transmitter release. If a drug acts presynaptically to inhibit transmitter release, then it will decrease the amplitude of the postsynaptic response but not affect the postsynaptic response evoked by application of exogenous transmitter. Ideally in such an experiment, one would wish to mimic exactly the amplitude and time course of the postsynaptic response by close iontophoretic or pressure application of exogenous neurotransmitter from fine micropipettes, which are positioned close to the cell under study.

A complication arises when the same inhibitory receptors are present on both the pre- and postsynaptic membranes. Examples of this are found with receptors which belong to the inhibitory, G-protein-coupled superfamily, i.e. α_2-adrenoceptor, M_2 muscarinic, D_2-dopamine, 5-HT_{1A}, $GABA_B$, A_1-adenosine, somatostatin, and μ- and δ-opioid receptors. When present on the postsynaptic membrane, activation of these receptors causes membrane hyperpolarization and inhibition of neuronal firing by increasing the membrane permeability to potassium ions. Would not such an inhibitory, postsynaptic effect by itself decrease or 'shunt out' the fast EPSP and thus make it difficult to examine any additional presynaptic inhibition which might be occurring? The answer is no. The current flowing during a fast EPSP is largely capacitative and, therefore, not attenuated by any increase in the conductance of the postsynaptic membrane (14). However, the time course of the EPSP would be altered by the change in postsynaptic conductance, becoming faster in onset and offset, since the membrane time constant is decreased.

4. Conclusions

With the techniques now available, detailed pharmacological and biophysical analysis of synaptic transmission in the CNS can be undertaken. The situation is obviously more complex than that previously examined at peripheral neuromuscular and ganglionic synapses. There is still a great deal to be understood.

References

1. Dingledine, R. (ed.) (1984). *Brain Slices*. Plenum Press, NY.
2. Madison, D. V. (1991). *Cellular Neurobiology: A Practical Approach*. (ed. J. Chad and H. Wheal). Oxford University Press, New York.
3. Blake, J. F., Brown, M. W., and Collingridge, G. L. (1988). *Neurosci. Lett.*, **89**, 182.
4. Finkel, A. (1991). *Molecular Neurobiology: A Practical Approach*. (ed. J. Chad and H. Wheal). Oxford University Press, New York.
5. Martin, A. R. (1955). *J. Physiol.*, **130**, 114.
6. Silver, R. A., Traynelis, S. F., and Cull-Candy, S. G. (1992). *Nature*, **355**, 163.
7. Davies, S. N. and Collingridge, G. L. (1989). *Proc. R. Soc. Lond. B*, **236**, 373.
8. Cherubini, E., North, R. A., and Williams, J. T. (1988). *J. Physiol.*, **406**, 431.
9. Egan, T. M., Henderson, G., North, R. A., and Williams, J. T. (1983). *J. Physiol.*, **345**, 477.
10. North, R. A. (1989). *Br. J. Pharmacol.*, **98**, 13.
11. Katz, B. (1966). *Nerve, Muscle and Synapse*. McGraw Hill, New York.
12. Edwards, F. A., Konnerth, A., and Sakmann, B. (1990). *J. Physiol.*, **430**, 213.
13. Larkman, A., Stratford, K., and Jack, J. (1991). *Nature*, **350**, 344.
14. Edwards, F. R., Hirst, G. D. S., and Silinsky, E. M. (1976). *J. Physiol.*, **259**, 647.

6

Quantal analysis of synaptic transmission

BRUCE WALMSLEY

1. Introduction

Our understanding of the mechanisms underlying synaptic transmission owes much to the early measurements and analysis of evoked and spontaneous synaptic potentials at neuromuscular junctions (1). There, two important observations were made:

- small, spontaneously occurring potentials were observed, whose peak amplitudes could be described by a unimodal distribution
- measurements of the nerve-evoked endplate potential amplitudes produced a multimodal distribution in which the peak occurred in integral multiples of the mean amplitude of the spontaneous miniature potentials (see *Figure 1*)

Thus, it was proposed that the spontaneous miniature potential represents a basic building block, or quantum, from which the evoked (multiquantal) endplate potential is composed. Following the electron-microscopic observation of membrane-bound vesicles within the presynaptic nerve terminal, it was further proposed that a spontaneous miniature endplate potential represents the postsynaptic response to the presynaptic release of the neurotransmitter contents of a single vesicle (i.e. a quantum of neurotransmitter). Release is thought to occur by fusion of the vesicle and presynaptic terminal membranes at specialized regions called active zones or release sites. Because of the existence of multiple quantal peaks in the histogram of evoked endplate potential amplitudes (see *Figure 1*), vesicular release of transmitter is assumed to be intermittent at each release site following the arrival of a presynaptic action potential, i.e. the release of the contents of a vesicle at a release site occurs in a probabilistic manner. The nerve-evoked postsynaptic potential then represents the sum of such probabilistic release at all release sites in the connection. It was proposed that the probability of release, p, may be the same at all release sites, and that release occurs from a total pool of 'available' vesicles, N. This formed the

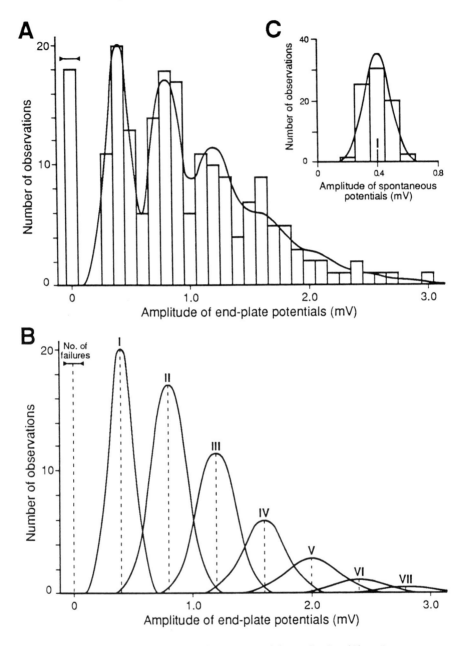

Figure 1. Histograms of evoked endplate potential amplitudes (A) and spontaneous endplate potentials (C) recorded from a cat neuromuscular junction. (B) shows a Poisson distribution of multiquantal components (I–VII, incorporating quantal variance) which are summed to produce the proposed fit (smooth curve in A) to the amplitude histogram. (Redrawn from reference 1, with permission.)

basis of the first structural models of quantal transmission in which release was described by either a Poisson or uniform binomial process (1, 2).

Since its introduction, further studies at neuromuscular junctions and at peripheral and central synaptic connections have led to either the confirmation, modification, or complete rejection of this model (for reviews see references 1–9). Most of these studies have been made using detailed measurements and analysis of the amplitude fluctuations of postsynaptic potentials and currents. This chapter summarizes the application and interpretation of a variety of models used in the fluctuation analysis of synaptic transmission.

2. Why use quantal models?

A model is used because we are unable to obtain complete and accurate measurements of the process being investigated. There are two major objectives in using quantal models of synaptic transmission:

(a) To provide a complete quantitative description of transmitter release and its postsynaptic action at the connection between two cells. It is usually hoped that such a description will provide some insight into the mechanisms underlying synaptic transmission, its regulation, and its structural basis at the connection being studied.

(b) To provide insight into the involvement of presynaptic and/or postsynaptic changes following modification of synaptic transmission. Such information is valuable in understanding the actions of neuromodulators or drugs, and the mechanisms underlying phenomena such as facilitation, post-tetanic potentiation, and long-term potentiation.

3. Structural basis and interpretation of quantal models

All models of synaptic transmission are based on or interpreted in terms of various structural (and ultrastructural) features of a synaptic connection, and a knowledge of these ultrastructural features is useful, if not essential, to the development of an appropriate model.

3.1 What is a release site?

There is a considerable diversity in the structural (and ultrastructural) features of synapses in both the peripheral and central nervous systems. At the neuromuscular junction of skeletal muscle, release is thought to occur from distinct regions of the presynaptic membrane exhibiting high concentrations of intramembranous particles, possibly Ca^{2+} channels, opposite the synaptic folds. Structural evidence supporting this comes from electron-microscopic observations of vesicle fusion with the presynaptic membrane and a clustering of vesicles adjacent to the presynaptic membrane in this region.

At many central (and some peripheral) synapses, a more highly structured presynaptic region is often observed and is thought to constitute a transmitter release site. In transmission electron microscopy of conventionally fixed tissue, this region has been observed as an obvious electron-dense band adjacent to the presynaptic membrane. The presynaptic density often appears as a regular array of protrusions under the electron microscope and at such synapses has been called the presynaptic grid. Vesicles are often found clustered at the presynaptic grid.

The ultrastructural appearance of a synaptic contact and the localization of synaptic vesicles in conventionally fixed tissue most probably provides a highly distorted or artefactual picture of this region. Recent electron microscopic studies of rapid-freeze material show a quite different picture of the presynaptic region with an array of filamentous material extending from the presynaptic membrane into the terminal, in which the synaptic vesicles appear to be bound (10, 11).

Vesicles tend to be more spherical in shape and exhibit considerably less variability in size in rapid-freeze material compared with conventionally fixed material (12). This is important because variability in the size of synaptic vesicles has been used in some studies as an indication of variability in the amount of transmitter released from those vesicles.

A synaptic cleft separates the pre- and postsynaptic membranes, and this synaptic cleft contains an increased density of filamentous material. The postsynaptic region immediately adjacent to the presynaptic density also exhibits a band of electron-dense material called the postsynaptic density. It is in this region that the postsynaptic receptors are thought to be localized in high concentration. The region defined by the pre- and postsynaptic densities separated by a synaptic cleft is called a synaptic specialization. Although there may be a large region of close apposition between the pre- and postsynaptic membranes, it has been proposed that vesicular release of transmitter only occurs at these specializations.

The presynaptic boutons at some connections may contain only a single specialization (5), or at other connections may contain multiple synaptic specializations (8). The size (area) and shape of the synaptic specializations is not uniform, but may vary considerably between boutons, and between specializations contained within the same bouton (8).

At many other synapses (e.g. some sympathetic varicosities), highly structured synaptic specializations are not observed under the electron microscope, although there may be regions of close apposition between the pre- and postsynaptic membranes (6, 13).

Although ultrastructural studies have provided much evidence regarding the identification of possible sites of transmitter release, further ultrastructural studies are required to determine whether transmitter release is strictly confined to these regions, and if so, to provide evidence on the nature and location of release *within* these regions.

3.2 What is a quantum?

The term quantum is commonly used to describe several aspects of synaptic transmission. It is most often used to describe the number of molecules of transmitter released as a 'packet' or 'quantum' from the presynaptic terminal. Under the vesicle hypothesis, a quantum corresponds to the amount of transmitter packaged into a single vesicle. It is also used to describe an incremental (or quantal) postsynaptic potential (or current) such as the spontaneous miniature endplate potentials at the neuromuscular junction. However, there is not necessarily a close correspondence between the quantal release of neurotransmitter and the observation of quantal postsynaptic potentials (or currents). Because of the technical difficulties in measuring transmitter release directly, measurements of synaptic transmission are most often made using recordings of the postsynaptic potential (or current), and a variety of interpretations of the structural basis of such measurements have been proposed. The released transmitter diffuses across the synaptic cleft and binds to receptors in the postsynaptic membrane; this results in the opening of the associated ion channels. Knowledge of the extent of the region occupied by receptors and the density of receptors in this region is important for the interpretation of physiological recordings, but unfortunately this information is often lacking. Most information on receptor localization has come from studies at neuromuscular junctions using acetylcholine as a transmitter.

Receptor-binding studies at some central synapses have shown that the receptor region is co-extensive with, and may extend beyond, the region defined by the postsynaptic density under the electron microscope (e.g. reference 14). Significantly, the area of this receptor region may vary considerably between specializations at the same connection.

The postsynaptic potential (or current) resulting from the presynaptic release of transmitter depends on the number of receptor–channel complexes activated. The number of receptor-channel complexes activated depends on many factors. An important factor is the nature of the diffusion of transmitter, which in turn depends on the geometry and structure of the synaptic cleft. Other important factors are the binding co-operativity of neurotransmitter at the receptors (i.e. the number of bound transmitter molecules required to activate the receptor), the breakdown and uptake of neurotransmitter following release, and desensitization of receptors. Theoretical analyses of diffusion and binding at a neuromuscular junction (15) led to the saturated disc model, in which a high density of receptors is activated following transmitter release out to an effective radius from the site of release, beyond which the density of activated receptors declines rapidly. The extent of the saturated receptor region depends on many factors, including the ratio of binding time to diffusion time (15). In this model, the amplitude of the postsynaptic potential (or current) is directly related (although not necessarily linearly) to the amount of neurotransmitter released. That is, the quantal nature of the postsynaptic response is directly a result of the quantal release of neurotransmitter. If more transmitter is contained in and released from a vesicle, then the area and number of activated receptors will be larger and the postsynaptic potential (or current) will be greater. Variation in the amount of neurotransmitter contained in vesicles will thus be reflected in the variability of amplitude of the quantal postsynaptic potentials (or currents). In addition, differences in the density of receptors between release sites at the same connection may result in different average quantal postsynaptic potentials or currents generated at different release sites. This source of variability is important to the interpretation of spontaneous quantal release (see Section 5.5).

An alternative hypothesis at a particular central synapse (16, 17) suggests that enough transmitter may be released from even a single presynaptic vesicle to saturate *all* of the postsynaptic receptors. Under this hypothesis, the quantal postsynaptic potential (or current) represents the (fixed) total number of available receptor–channel complexes and is insensitive to the amount of transmitter released by each vesicle (or from the simultaneous release of transmitter from several vesicles at the same release site). In this case, the quantal nature of the postsynaptic potential (or current) has a primarily *postsynaptic structural basis*. Differences in the density and number of receptors between release sites may comprise a major source of variability in the quantal postsynaptic response, under this hypothesis (see Section 5.5).

A recent study of an inhibitory synapse in the hippocampus has suggested that the observed quantal postsynaptic current is due to the activation of a *cluster* of receptor–channel complexes of a tightly regulated number (18). Such a cluster could be an example of the saturated receptor model (16, 17), or there could be a number of such clusters at a single postsynaptic site. Following the observation of subminiature endplate potentials at the frog neuromuscular junction (19), it has been proposed that a 'quantal' miniature endplate potential is actually composed of the sum of a number of subminiature endplate potentials, each corresponding to release of the neurotransmitter contents of a single vesicle (19).

Although there are a variety of other hypotheses, including non-vesicular release models, these examples indicate that some hypotheses of the fluctuations in amplitude of synaptic potentials (currents) relate the 'quantal' nature of these fluctuations to the presynaptic release of quanta of neurotransmitter and some relate them directly to the properties of the postsynaptic receptor–channel region. Obviously, the interpretation of quantal analysis of synaptic potentials (or currents) therefore depends on the chosen structural model.

3.3 What is N?

Under the vesicle hypothesis of quantal transmission, there exists a total population, N, of vesicles 'available' for release. The more restrictive proposal, made at some connections which exhibit distinct synaptic specializations, is that N corresponds to the total (physical) number of these specializations (presumed release sites; 2, 5, 8). Several interpretations have been made concerning this proposal. The first equates a quantal postsynaptic potential with the presynaptic release of a single vesicle. Under this hypothesis, only a single vesicle (or none) is released at each release site following a presynaptic action potential. An important issue with this model is whether multiquantal release at a single release site is absolutely 'forbidden' by some physical mechanism, or is simply a highly improbable occurrence under the conditions examined. This is important to the interpretation of the results of experiments in which release is altered. A second hypothesis allows multiquantal release at a release site, but relates the quantal postsynaptic potential to the properties of the postsynaptic receptor region (the saturated receptor model; 16, 17). In both cases, the maximum number of quantal components of the postsynaptic potential (or current) is equal to the number of transmitter release sites. However, the structural interpretations are obviously quite different.

Under some conditions, release may not occur at some of the release sites in a connection. In this case, the experimenter must carefully consider the *definition* of the total population of 'release sites' as corresponding to the physical (anatomical) number of release sites or to the number of active or 'working' release sites (20, see compound binomial quantal model, Section 5.4.2.*ii*.).

Other models equate N with the number of 'available' vesicles (allowing multivesicular release at a single release site), the number of vesicle clusters, the number of presynaptic transmitter pores, the number of postsynaptic receptor clusters (allowing for the possibility of several clusters at the same release site), or to some combination such as release site plus available vesicles (21).

These examples illustrate that, even under the vesicle hypothesis of transmitter release, there may be a variety of structural interpretations that are compatible with the results of quantal analysis of synaptic potentials. The same data may also be compatible with a variety of other structural models of transmitter release, and it is therefore important that experiments be designed to rigorously test the validity of a proposed model.

A knowledge of the structural (and ultrastructural) features of a synaptic connection is clearly useful (although not necessarily conclusive) in arguments related to the physical correlate of N, determined from measurements of quantal synaptic potentials. Obtaining this structural information is obviously technically demanding at most connections. This is particularly true if there are multiple presynaptic cells (e.g. central neurones), where intracellular labelling of the presynaptic (and probably postsynaptic) neurone must be used to identify putative contact regions (5, 22–25). Serial section electron microscopy of all putative contact regions is then required to reveal (confirm) and count the total number of specializations, and establish their geometry (a procedure not without error). Although complete structural information may not be obtained, any information on the location and number of synaptic contacts is likely to be useful in the interpretation of the time course and amplitude fluctuations of synaptic potentials (and currents).

4. Recording and measurement of synaptic potentials and currents

Obtaining stable and relatively noise-free intracellular recordings of synaptic potentials and currents is by far the most important step towards a reliable analysis of the fluctuations of these potentials and currents. Unfortunately, intracellular recordings are always associated with noise, which contaminates the measurements of both spontaneous and evoked synaptic potentials and currents. One source of noise is due to the electrode resistance, which should be kept to a minimum (26). The use of whole-cell patch electrodes may be valuable for this purpose (27). The other source of noise is due to the opening and closing of membrane channels, including that from ongoing synaptic activity in the cell. All attempts should be made to reduce ongoing synaptic activity, not only because it reduces the background noise, but also because it minimizes the possible complicating effects of non-linear interactions between the measured synaptic potentials and the background synaptic activity (28).

An important aspect of the subsequent analysis is the measurement and characterization of the contaminating noise (see Section 5.2). A second limitation usually associated with intracellular recordings is the length of time during which stable intracellular measurements can be made. This restricts the number of measurements of evoked and spontaneous events that can be obtained, and this means that statistical sampling theory must be incorporated into the subsequent analysis (see Section 5).

4.1 Synaptic potentials or currents?

An important choice is whether to make voltage or current measurements of the postsynaptic response. Each has advantages and disadvantages.

The signal-to-noise ratio may be much better for voltage recordings than for voltage-clamp measurements of synaptic currents, using discontinuous or switched single-electrode voltage-clamping with conventional microelectrodes (27). However, whole-cell patch-clamp recordings of synaptic currents may provide significant improvement in the signal-to-noise ratio if this technique can be successfully applied.

The problems of obtaining an adequate space-clamp should be addressed in voltage-clamp recordings, since an inadequate space-clamp can lead to distortions of the synaptic current records (27).

A possible complication associated with synaptic potential recordings is the existence of a nonlinear synaptic current–voltage relationship due to a reduction in the driving potential for the synaptic current, caused by the synaptic potential itself. The severity of this effect depends on the relative magnitude of the synaptic potentials compared with the reversal potential for the synaptic current, and on the time course of the synaptic conductance transient compared with the membrane time constant of the cell. Correction for this non-linearity is not trivial, and depends on a determination of the electrotonic properties of the cell and the electrotonic locations of the synaptic contacts. This is important because the non-linearity is generated at the synaptic contacts, where the amplitude of a synaptic potential may be considerably greater than at the recording electrode. A further complication arises if the time course of the synaptic conductance change is also voltage-dependent, and this is often the case. In cases where the synaptic contacts are electrotonically close to the recording electrode, correction formulae for non-linearities have been used (2, 29, 30). In general, however, if the effects of non-linearity are thought to be significant, it may be preferable to use voltage-clamping of the synaptic current to avoid this problem.

4.2 Automatic baseline correction

In addition to minimizing the effects of contaminating noise due to electrode resistance and membrane noise, it is equally important to minimize the noise, loss of resolution, and distortions which might occur during recordings of the signals onto tape or directly onto computer. Since the synaptic potentials and currents are often small, this usually necessitates the use of a high-gain DC recording.

A useful technique in maintaining high-gain recordings of evoked potentials and currents is to use an automatic baseline stabilization procedure, either using analog or digital circuitry as a sample-and-hold device (16). At the beginning of each oscilloscope sweep, the membrane potential is sampled briefly, and stored in the sample-and-hold device or computer. This value is then subtracted from the signal during the subsequent recording period, usually by applying the output of the sample-and-hold device in 'hold' mode to the negative input of a differential amplifier (with the signal on the positive input). Because the signal is then automatically set to zero at the beginning of each sweep (i.e. no DC offset), a high-gain DC amplification can be achieved using this procedure. Care must be taken, however, in choosing a sample-and-hold device or digital-to-analog (D-A) computer output which does not introduce additional noise or drift (sag) during the recording period.

4.3 Initial analysis of evoked responses and noise contamination: choosing baseline and measurement regions

Having obtained a series of recordings of evoked potentials (currents), the initial step in fluctuation analysis is to measure the amplitudes of these potentials (currents). These measurements are performed by choosing a baseline region before the onset of the evoked potential (current) and a measurement region during the evoked potential (current) or noise record (see *Figure 2A*). The average value over the baseline region is calculated and subtracted from the average value over the measurement region. These calculations are performed for each response. The choice of baseline and

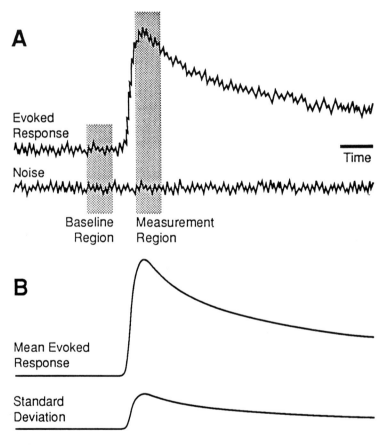

Figure 2. Schematic diagram illustrating the measurements of single evoked responses and noise records (A), and the mean and standard deviation time courses of the evoked responses (B). See text for further explanation.

measurement regions is an extremely important step in minimizing the effects of the contaminating noise; these regions should be chosen with care. Several points, covered in the following sections, should be considered when choosing the baseline and measurement regions.

4.3.1 Peak amplitude or time integral?

Both peak-amplitude and time-integral measurements have been used to analyse amplitude fluctuations in postsynaptic potentials and currents. In practice, an optimum signal-to-noise ratio can be achieved by choosing a measurement region around the peak amplitude which corresponds to a finite time window following the stimulus (see *Figure 2A*), and calculating the mean amplitude of the response over this region. A time window immediately prior to the beginning of the evoked response should be chosen as a baseline region (carefully avoiding stimulus and field potential artefacts). Measurements of the contaminating noise are made using these *same* time windows in the absence of stimulation, as illustrated in *Figure 2A*. The noise recordings should be obtained between successive evoked responses, so that the same recording period is used for both signal and noise. Because there are no restrictions on the length of the noise recordings, many more noise measurements can be made than evoked response measurements. This is advantageous in reliably characterizing the properties of the noise for the subsequent fluctuation analysis.

There are a variety of factors which determine the optimum time windows for the evoked response and the noise, and an empirical approach is often used to find the optimum regions, i.e. a range of different windows is tested and the one producing the best signal-to-noise ratio is chosen. The duration of the optimum baseline and measurement windows may be different (e.g. long-duration baseline region and short-duration measurement window).

The definition of signal-to-noise ratio in the context of fluctuation analysis requires careful consideration. The simplest approach is to calculate the average amplitude of all of the evoked responses (for the chosen baseline and measurement regions) and divide this value by the standard deviation of the noise amplitude measurements (for the same time windows). Baseline and measurement regions are then found empirically which maximize the ratio of the average evoked-response amplitude to the noise standard deviation.

Since we are interested in amplitude *fluctuations* and not in the average amplitude of the evoked response *per se*, another approach is to maximize the ratio of the variance (or standard deviation) of the evoked-response amplitudes to the variance (or standard deviation) of the noise amplitudes. However, in practice this approach may lead to the selection of a measurement region of the evoked response which is not representative of amplitude fluctuations, but rather some other highly variable aspect of the evoked response, such as latency fluctuations (see below).

As a general guideline, measurements on the rising phase of the evoked response are sensitive to this problem and should be avoided, and the measurement region should straddle the peak of the evoked response. The baseline region should be as close as possible to the beginning of the evoked response, and the duration of this window chosen to maximize the signal-to-noise ratio.

The choice of a finite time window results in response amplitude measurements that are neither true peak measurements nor complete time-integral measurements. A simple form factor to correct these values to, say, peak amplitude measurements, can be obtained by dividing the peak amplitude of the averaged population response by the mean value over the chosen time window for the population average. This correction factor relies on the time course of all fluctuation components being identical and will not be correct if there are significant latency fluctuations or other influences such as electrotonic effects.

4.3.2 Latency fluctuations

The contribution of each (quantal) component of the evoked response does not necessarily occur at the same latency and with the same time course following the arrival of an action potential. For example, variable latency of the release of quantal components may occur, and the arrival of an action potential at all of the synaptic terminals in a connection may not be simultaneous (31–33). These factors may cause significant latency fluctuations to occur in the evoked response and severely affect the amplitude measurements.

An indication of significant latency fluctuations may be obtained by examination of the standard deviation time course of the evoked response (16). If the evoked response is $y(t)$ and the noise is $n(t)$, where t(time) is usually represented as a discrete series of successive D-A sample times, then the standard deviation time course of $y(t)$ about its mean value $\overline{y(t)}$ is:

$$\left\{ \frac{1}{(N-1)} \sum_{i=1}^{N} [y_i(t) - \overline{y(t)}]^2 - \frac{1}{(M-1)} \sum_{i=1}^{M} [n_i(t) - \overline{n(t)}]^2 \right\}^{1/2}$$

where N denotes the number of evoked response records and M is the total number of noise records. Because of finite sampling the variance subtraction can assume negative values on occasion, e.g. during the baseline. These points can be excluded or plotted for convenience as negative values (having taken the square root of the modulus of the variance value, (16)). If each fluctuation component contributing to the evoked response is identical in time course and latency, then the standard deviation time course should be the same as the time course of the averaged response, i.e. it should have a peak amplitude at the same time as the average response (see *Figure 2B*). Significant latency fluctuations are evident as a peak response in the standard deviation time course during the rising phase of the average response. In practical terms, measurements during the rising phase of the evoked response will be most affected by latency fluctuations and should be avoided. Measurement of the time integral of the evoked response over the entire time course will produce an amplitude measurement insensitive to latency fluctuations. However, this measurement may produce an extremely poor signal-to-noise ratio which precludes subsequent fluctuation analysis.

If the standard deviation time course is not the same as the averaged response, then this could also be an indication of differences in the *time course* of the underlying fluctuation components. An important factor contributing to these time course differences is the effect of the electrotonic properties of the postsynaptic cell.

4.3.3 Electrotonic effects

Most synaptic connections involve multiple contacts between the presynaptic cell and the postsynaptic cell. This inevitably results in different distances between these contacts and the recording electrode. For example, the same presynaptic axon may give rise to some contacts on the soma of a neurone and some along the dendrites. The electrotonic membrane properties of the postsynaptic cell may attenuate the amplitude and distort the time course of the synaptic potentials (and currents) as they propagate towards the recording electrode (33, 34).

The severity of electrotonic effects depends on many factors, such as the time course of the synaptic conductance change relative to the membrane time constant and the electrotonic structure of the cell (e.g. the electrotonic lengths of dendritic branches). A knowledge of the anatomical details of the connection is also clearly important to the interpretation of this issue. Calculating the membrane properties of a neurone and its dendritic branches from measurements made with a somatic recording electrode is extremely difficult, and often complicated by the existence of non-uniform and non-linear membrane properties. However, this issue must be addressed in any interpretation of the amplitude fluctuations of synaptic potentials (or currents); a more detailed discussion and references can be found in references 33 and 34.

The following points are relevant to fluctuation analysis:

(a) It is *not* valid in general to use the time course of a synaptic potential in conjunction with an electrotonic model of the neurone to verify a restricted electrotonic location for the synaptic contacts, although in some cases spatial dispersion may be detected (33).

(b) One way to avoid the problem of electrotonic distortions is to restrict fluctuation analysis to synaptic connections that are close to the recording site. However, this region is usually where the *worst* (steepest) electrotonic distortion and attenuation with distance from the recording site occurs (33, 34). Unless the synaptic contacts are actually *at* the recording site (e.g. the soma), electrotonic effects due to even small distances between synapses may be significant and an estimation of these effects should be made.

(c) Most synaptic connections are not made conveniently at the soma, and in any case it is often required to analyse the properties of synaptic potentials generated on the dendrites of a neurone. Although the amplitude and time course of the synaptic current arising from dendritic synapses is attenuated and distorted, it may be possible to analyse the properties of the fluctuations of the postsynaptic potentials if it can be shown that they are generated at similar electrotonic locations (but see above). In contrast to juxtasomatic contacts, incremental changes in the attenuation of synaptic potentials (measured at the soma) with distance may be small for synaptic connections on distal dendrites (33, 34).

(d) Anatomical evidence on the location of synaptic contacts is obviously valuable, but not necessarily conclusive, in arguments related to electrotonic effects. Although all of the synaptic contacts may be identified anatomically, the release properties of each contact are not known and it may not be appropriate to assume that release is the same (or occurs at all) for all of the identified contacts. However, anatomical studies are obviously useful in revealing the sites of possible release and providing a basis for estimating electrotonic effects (23, 24).

4.4 Spontaneous potentials and currents

Incorporation of measurements of spontaneous potentials and currents in quantal models is described in detail in Section 5.5. However, the following points should be noted:

- ideally, measurements of spontaneous potentials and currents should be made with the same baseline and measurement regions used for the evoked responses, to allow these measurements to be used directly in the subsequent quantal analysis (or the measurements appropriately corrected)
- reliable measurements of contaminating noise should be made to determine the true (noise-free) spontaneous potential (current) distribution (see Section 5.5)

There are many potential sources of error and/or misinterpretation in the recording of spontaneous events, including:

- detection bias (small spontaneous events may not be detected due to the background noise)
- noise bias (noise measurements may be contaminated by spontaneous events, i.e. non-independence of signal and noise)

- the source(s) of spontaneous events may be unknown, particularly if there are multiple presynaptic cells (but see reference 30)
- electrotonic effects may attenuate and distort spontaneous events (see Section 4.3.3)
- amplitude variability may be due to variability at the same release site or differences between release sites. These two sources of variability cannot be treated as a common quantal variance in the subsequent analysis (see Section 5.5)

5. Models of amplitude fluctuations of synaptic potentials (currents)

The starting point for fluctuation analysis is a string of amplitude measurements of the evoked synaptic potential (current), a corresponding string of noise measurements, and in some cases a string of measurements of spontaneous potentials (currents). The object of further analysis is to use these sampled data as efficiently as possible to reveal information about the true nature of the evoked and/or spontaneous potential (current) fluctuations.

Because only limited samples of the data are available, statistical methods must be used in this analysis. This usually involves proposing a particular model and subsequently determining the optimum values of the unknown parameters of that model. One approach to this problem is to use the *method of moments*, in which the sample moment serves as an estimate of the corresponding population moment (e.g. mean and variance of a normal distribution). This method has been used extensively in quantal analysis in relation to the Poisson and simple binomial models (1, 2). Another approach is to use an optimization procedure to determine the parameters of a model according to some criterion of goodness-of-fit of the proposed model to the measured data. The most commonly used criteria to fit the data are minimizing the squared-error, minimizing the Chi-squared statistic and maximizing the likelihood function. Optimization procedures which minimize the squared-error or Chi-squared statistic require binned data (e.g. histograms of the synaptic potential amplitudes), whereas the maximum likelihood method usually operates on the raw (unbinned) data, e.g. string of amplitudes of the synaptic potential, without the loss of information suffered as a consequence of binning the data (although the maximum likelihood method may also be applied to binned data).

Both probability theory and simulation studies with sampled data demonstrate that the *maximum likelihood estimator* (MLE) provides the most consistent and reliable estimator of the parameters of a variety of statistical models, and this approach is therefore the one recommended for use in fitting quantal models (35–39).

5.1 Maximum likelihood method

The basis of the maximum likelihood approach to fitting a hypothesis (model) to the data is as follows: given the data (e.g. a string of measurements of synaptic potential amplitudes), the question 'What is the plausibility or likelihood that the hypothesis (model), if correct, has produced this data?' is asked. Since the data is given, the optimum parameters of the model are those which maximize the likelihood of observing the data.

Consider a string of m measurements of synaptic potential amplitudes ($v_1, v_2 \ldots v_m$). We wish to fit this sampled data with a chosen model which describes the complete distribution of synaptic potential amplitudes (v) with a probability density function $f(v|\phi)$ where (ϕ) represents the unknown parameters of the mode ($\phi_1, \phi_2 \ldots \phi_k$). For example, a Gaussian or normal distribution of amplitudes would be described by the probability density function

$$f(v \mid \phi) = \frac{1}{\sqrt{2\pi\sigma^2}} e^{-(v-\mu)^2/2\sigma^2}$$

where the unknown parameters (ϕ) = (ϕ_1, ϕ_2) are μ and σ, the mean and standard deviation, respectively.

The likelihood of observing all of the measured data values ($v_1, v_2 \ldots v_m$) with a given set of parameters ($\phi_1, \phi_2 \ldots \phi_k$) is proportional to the product of the probability densities $f(V \mid \phi)$ evaluated at all of the individual data values:

$$\text{Likelihood } (\phi) = f(v_1 \mid \phi) f(v_2 \mid \phi) \ldots f(v_m \mid \phi).$$

This can be simplified by taking the logarithm of this product to produce a sum:

$$L(\phi) = \log \text{Likelihood } (\phi) = \sum_{i=1}^{m} \log f(v_i \mid \phi).$$

The particular (optimum) values of the parameters ($\phi = \phi_1, \phi_2 \ldots \phi_k$) which maximize the Likelihood (ϕ) are the same as those which maximize the log Likelihood function $L(\phi)$, and are called the *maximum likelihood estimates* of these parameters (35–38).

A necessary condition for a maximum of the log-likelihood function is met where the first partial derivative of this function with respect to all (ϕ) is equal to zero:

$$\frac{d \log L (\phi)}{d \phi_i} = 0 \text{ for } i = 1, 2 \ldots k.$$

(A sufficient condition for this to be a maximum, rather than a minimum or point of inflexion is that the second partial derivatives are negative.)

Except for a number of special cases, explicit solutions for this system of equations (which may have multiple and/or complex roots) are not available, and *iterative methods* must be used to find a realistic global maximum. A variety of iterative methods have been applied to this problem (see references 35, 36, 38–40). Two main techniques have been used in fluctuation analysis of synaptic potentials (and currents):

(a) *Constrained-gradient or direct-search method.* This procedure uses a particular method (Simplex, Newton–Raphson, Davidon–Fletcher–Powell, see 35, 41, 42) to search the log-likelihood function directly for a global maximum, subject to the constraints of the model. These methods have several problems, one of which is that convergence to a global maximum, rather than a local maximum (there may be several maxima) may depend strongly on the initial estimates of the parameter values of the model. Solution to this problem then depends on either accurately estimating the initial parameter values, or calculating the solutions for a wide range of possible initial parameter values. Detailed descriptions of this method can be found in (35, 41–43).

(b) *E–M Algorithm.* An iterative solution to finding the maximum likelihood estimates of the parameters of a proposed distribution, recognizing that the data is an *incomplete* sample from that distribution, is the E–M (expectation–maximization) algorithm (35, 36, 38). Each interation proceeds with expectation step followed by a maximization step. Starting with an initial estimate of the unknown parameters ($\phi^{(o)}$), each iteration generates a new estimate of these parameters ($\phi^{(m+1)}$) from the previous estimate ($\phi^{(m)}$). The E–M algorithm has generally been found to be less sensitive to initial parameter estimates and more accurate than search methods in a variety of simulation studies (35). A good description of the E–M algorithm with examples is provided by Everitt (35; see also reference 36 for computer program listing). Kullmann (40) provides a description of the E–M algorithm adapted for use in quantal analysis (see also reference 39). The E–M algorithm is generally the maximum likelihood method of choice for fluctuation analysis, because of its robustness, and the recognition and incorporation of the measured data as an incomplete sample.

5.1.1 Confidence intervals and goodness-of-fit tests

An indication of confidence intervals for quantal models has usually been obtained by the use of *Monte-Carlo simulations* and these should be performed or consulted for the particular model/experimental condition examined (39, 40, 44, 45). However, such simulations do not indicate how much better the chosen solution is to solutions at other possible local maxima, nor do they indicate that the chosen solution (or model) is the correct one.

Multiple samples of the data should be examined for consistency (e.g. as histograms). In addition, the maximum likelihood results for these samples should be plotted and used to indicate confidence bounds (see references 7 and 40).

Acceptance or rejection of a model following estimation of parameter values may also be made using conventional goodness-of-fit tests such as the Kolmogorov–Smirnov and Chi-squared tests, and comparison of alternative models may be made using Likelihood-ratio tests (37, 43, 46). *Acceptance* of a model by goodness-of-fit tests should be made cautiously, since a wide variety of different models may fit the same data equally well (47, 48). Although a fit is made using the individual data values, illustration of the fit is usually made by constructing a histogram of the data. However, a histogram is a completely different form to the continuous probability density function (PDF) of the model, and this provides a somewhat artificial comparison. An alternative, for illustrative purposes only, is to plot each data point as a triangle centred on the data value, whose base on the amplitude axis is extremely small, and to sum all of these triangles to produce a smooth distribution (see *Figures 3* and *4*).

The following sections describe the procedures for developing and applying particular models.

5.2 Noise contamination and deconvolution

Measurements of evoked and spontaneous potentials are always contaminated by background noise. This means that the observed fluctuations exhibit a greater variability than would be measured in a noise-free situation. If it is assumed that:

(a) the synaptic potential (current) and the noise add linearly
(b) the fluctuations in the amplitude of the synaptic potential (current) and the noise are statistically independent, then:

$$f(V) = \int_x S(x)N(V-x)\mathrm{d}x$$

where $f(V)$ is the PDF of the noise-contaminated synaptic potential, $N(V)$ is the PDF of the contaminating noise, $S(V)$ is the PDF of the noise-free synaptic potential, V is voltage, and x is the integration variable. $f(V)$ is a *convolution* (see reference 45 for example) of $S(V)$ and $N(V)$. Procedures used to find the noise-free synaptic potential distribution from the measured synaptic potential and noise distributions are generally called *deconvolution* procedures, all of which make further assumptions concerning the nature of the synaptic potential fluctuations and/or the contaminating noise. Assumptions (a) and (b) may not be correct under some circumstances, particularly if there is a considerable amount of background spontaneous synaptic activity (28).

It is essential to obtain accurate measurements of the contaminating noise to ensure reliability of the subsequent deconvolution procedures. Although there may be experimental limitations on the number of evoked responses which can be recorded, this limitation does not usually apply to the number of noise samples which can be measured. A number of different approximations to the (continuous) noise PDF have been used. In some cases, the noise measurements can be fitted by a normal or Gaussian distribution (40; see also 'Test for normality' in 46). However, in many cases, intracellularly recorded noise is not well described by a normal distribution and the noise amplitude distribution may exhibit considerable degrees of skew and/or kurtosis. In this case, the use of a normal approximation to the noise may produce serious errors in the subsequent deconvolution (40). Kullmann (40) has found that intracellular noise distributions can often be well modelled by a sum of two normal distributions, whose means, variances, and probabilities can be obtained using the E–M algorithm. If the noise cannot be simply described, and a very large number of noise samples has been obtained (usually many thousands of samples), a histogram or frequency polygon of the actual noise measurements may be used to approximate the continuous noise PDF (20, 49 see *Figure 3*).

Having obtained a reliable approximation to the noise PDF, the next step is to obtain maximum likelihood estimates of the parameters of the particular model(s) of interest.

5.3 Formulating a model

The convolution of the proposed model PDF and the noise PDF is as follows:

$$f(V|\phi) = \int_x S(x|\phi)N(V-x)dx$$

where $S(V|\phi)$ is the proposed noise-free PDF, (ϕ) represents the unknown parameters of the model ($\phi_1, \phi_2 \ldots \phi_k$), $N(V)$ is the noise PDF, and $f(V|\phi)$ is the noise-contaminated PDF (which also incorporates the unknown parameters (ϕ)). $f(V|\phi)$ represents the (noise-contaminated) PDF to be fitted to the measured (noise-contaminated) data ($v_1, v_2 \ldots v_m$) by calculating the maximum likelihood estimates of ($\phi_1, \phi_2 \ldots \phi_k$).

The proposed model $S(V|\phi)$ can be:

(a) a *continuous PDF*, e.g. a normal PDF of spontaneous potential amplitudes (see Section 5.5)

(b) a *discrete PDF*, e.g. a binomial quantal distribution, with negligible quantal variance (see Section 5.4)

(c) a *finite sum of continuous PDFs*, e.g. a binomial distribution of quantal components with variance (see Section 5.5)

With all of these models, it is obviously advantageous if an explicit expression can be obtained for $f(V|\phi)$, and this is usually the case if $N(V)$ is normally distributed, or can be well described by other continuous functions, such as the sum of two normal distributions (see Section 5.2). However, if an explicit expression cannot be obtained, $f(V|\phi)$ can be evaluated using numerical convolution for particular values of the unknown parameters (ϕ), during the optimization procedure (45).

The following sections describe the models most commonly used to analyse synaptic potential (current) amplitude fluctuations.

5.4 Discrete PDF models

These models describe the synaptic potential (current) amplitude distribution as a series of discrete amplitude components:

$$S(V|\phi) = \sum_{i=1}^{k} P_i \delta(V-V_i)$$

where k is the number of discrete components, δ is the delta function, and P_i is the probability of the ith component, subject to the constraints $0 \leq P_i \leq 1$ and $\Sigma P_i = 1$. In this case the unknown parameters (ϕ) are the k probabilities (P_i) and the associated k amplitudes (V_i) of the components.

The effect of convolving $\delta(V-V_i)$ with $N(V)$ is to simply shift $N(V)$ to $N(V-V_i)$. Therefore, for the discrete model:

$$f(V|\phi) = \sum_{i=1}^{k} P_i N(V-V_i) \text{ subject to } 0 \leq P_i \leq 1 \text{ and } \Sigma P_i = 1.$$

A special case occurs if the noise PDF, $N(V)$, is normally distributed with mean μ_N (usually zero) and variance σ^2_N:

$$f(V|\phi) = \sum_{i=1}^{k} \frac{P_i}{\sqrt{2\pi\sigma^2_N}} e^{-(V-\mu_i)^2/2\sigma^2_N} \quad \text{where } \mu_i = V_i + \mu_N.$$

In this case, the measured data is viewed as an incomplete sample from a mixture (sum) of normal distributions of unknown means (V_i) and with variances equal to the variance of the noise (σ^2_N). The E–M algorithm is ideally suited to this problem (35). Kullmann (40) has described the application of the E–M algorithm to this model, and to the case where the noise is modelled by the sum of two normal distributions.

A number of constraints may be placed on the discrete model to obtain more specific models. These will be considered in order of increasing constraints.

5.4.1 Non-quantal models (generalized deconvolution)

This model places no further constraints on $f(V \mid \phi)$, i.e.

$$f(V|\phi) = \sum_{i=1}^{k} P_i N(V-V_i) \text{ subject to } 0 \leq P_i \leq 1 \text{ and } \Sigma P_i = 1.$$

The rationale for this model is that a wide range of possible noise-free PDFs may be approximated by a series of discrete components. For example, a continuous distribution may be approximated by a series of closely spaced discrete amplitude components. In some cases the 'quantal' components underlying an evoked synaptic potential (current) may exhibit amplitudes which are discrete (constant amplitude at each release site), but may have different amplitudes at each release site. In this case, a discrete 'non-quantal' amplitude fluctuation pattern may result. Thus, the perceived advantage of this procedure is that the model is not constrained by any particular assumption, such as identical amplitude quantal components. However, in practice, the signal-to-noise ratio is such that this perceived advantage has not been realized and the following observations have been made concerning this model:

(a) The resolution of the technique depends on many factors such as the sample size, the number of discrete components (initially unknown), the separation of these components compared with the noise standard deviation, and the individual probabilities of these components (39, 40, 44).

(b) In general, components separated by less than 1.5–2.0 σ_N (noise standard deviation) *cannot* be resolved. (This condition is approximated when individual peaks in the convolved PDF are readily observed.) Unfortunately, this requirement severely limits this model to those cases in which the true amplitude fluctuation pattern is actually a series of widely separated (compared with σ_N) discrete amplitude components.

(c) The true underlying fluctuation pattern is not known *a priori* and, under *limiting resolution conditions*, artefactual solutions consisting of equally spaced (quantal) components may be obtained (44, 50). The considerable freedom of this model to produce excessively good fits to the data should be kept in mind when making comparisons with fits obtained using other more restricted models.

The generalized deconvolution technique has been used in a number of experimental studies and, in general, the results have been used to provide evidence either for or against an underlying discrete and/or quantal process. Examples of these results (see *Figure 3*) and further discussion of this issue

Quantal analysis of synaptic transmission

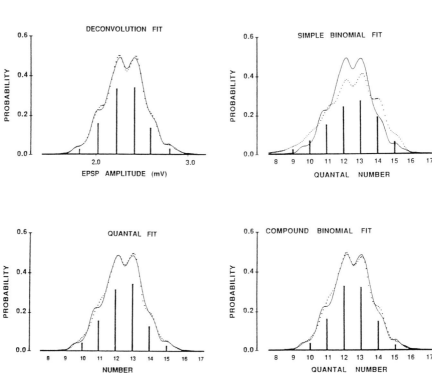

Figure 3. Top: Measured noise (scale = 0.006 μV^{-1}) and EPSP amplitude distributions, obtained in a cat spinal cord neurone (20). Below: Maximum likelihood fits (dotted lines) to measured synaptic potential amplitude distributions (smooth curves) for four different discrete model PDFs (solid bars). Probability scales apply only to proposed noise-free PDFs. See text for further explanation.

can be found in (7–9, 39, 44, 49, 50). Application of the E–M algorithm to this procedure is given in (40).

5.4.2 Quantal models

Quantal models are similar to the generalized deconvolution model, but further constrain the discrete amplitude components to be separated by an equal or *quantal* increment:

$$f(V|\phi) = \sum_{i=0}^{k} P_i N(V-iQ)$$

where k is the number of quantal components, Q is the quantal increment, $N(V)$ is the noise PDF, and P_i is the probability of the ith (multiquantal) component, subject to $0 \leq P_i \leq 1$, and $\Sigma P_i = 1$.

The assumption of an underlying quantal process may be based on a number of observations, such as the spontaneous postsynaptic potential (current) amplitude distribution (see Section 5.5). The occurrence of regularly separated peaks in the histogram of evoked synaptic potential (current) amplitudes is often taken as good evidence for an underlying 'quantal' process. However, the following points should be carefully considered if this is the sole basis for assuming a quantal process:

(a) An 'apparent' quantal distribution of synaptic potential (current) amplitudes may be generated by a non-quantal process consisting of the sum of a number of discrete components of quite different amplitudes. The spacing between the peaks is then not representative of quantal size, but instead reflects the difference between the discrete amplitudes of the underlying components.

(b) Small sample sizes should be avoided since sampling of even a smooth, continuous distribution can produce artefactual peaks (see examples and tests in reference 51). Multiple samples should be tested for consistency.

Because the generalized deconvolution model does not assume any particular discrete distribution, it is not relevant to include *quantal amplitude variability* in this model. However, quantal variability can be incorporated into models assuming an underlying quantal process, and this is considered in Section 5.5, following a description of three quantal models which assume that the quantal variability is negligible compared with the noise variance (which may be approximated at some synaptic connections (5, 8, 18).

i. Generalized quantal model

The generalized quantal model makes no further assumptions and simply invokes the quantal constraint. It makes no assumptions about the probabilistic relationship between successive multiquantal components, does not require stationarity in the values of the probabilities (P_i), and makes no prediction of a total population of available quanta. It does require stationarity in the quantal amplitude Q. In this case:

$$f(V|\phi) = \sum_{i=0}^{k} P_i N(V-iQ) \text{ subject to } 0 \leq P_i \leq 1 \text{ and } \Sigma P_i = 1$$

where k is the number of quantal components, Q is the quantal increment, and P_i is the probability of the ith (multiquantal) component. (P_0 is the probability of failures of response). A sufficient number of components k is chosen so that any further increase does not alter the solution (i.e. zero P for additional components).

Often there is no independent measurement of the quantal amplitude (e.g. from spontaneous potentials). In this case, the generalized quantal model may be useful in determining a likely value for the quantal amplitude, Q. This procedure has been used by Walmsley *et al.* (20) to examine the applicability of a quantal model, assuming that quantal variance is negligible. Maximum likelihood solutions are plotted and compared for a selected range of quantal amplitudes, and the optimum quantal amplitude chosen by inspection (see Figures 2 and 4 in reference 20). Although the quantal amplitude can be directly incorporated as an unknown parameter, it is useful to obtain and inspect such a plot for several reasons.

Firstly, there may be multiple maxima with similar fits to the data, but with different values of quantal size, Q. The appearance of multiple maxima is accentuated by the fact that a quantal distribution is constrained to the amplitude origin (20). Also, as the quantal size is allowed to become progressively smaller, 'identical' solutions will occur in which the quantal size is one-half that at a previous local maximum. It is clearly important to select an 'appropriate' range of quantal amplitudes to be tested, and this choice must be left to the experimenter (20).

The advantages of this model are that stationarity in the component probabilities is not required, nor is it necessary to assume a particular probabilistic quantal model. In addition, it is useful to compare the results with those obtained using more constrained quantal models (see Section 5.4.2.*ii*).

An example of a generalized quantal fit is illustrated in *Figure 3* (quantal fit). See reference 40 for the application of the E–M algorithm to this model.

ii. Stationary quantal models

The generalized deconvolution and generalized quantal models are designed to reveal the pattern of amplitude fluctuations of synaptic potentials (currents) from data obtained over a certain time period. Although these procedures are usually applied to stationary data (no change in parameters with time), stationarity in the probabilities of the fluctuation components is not a necessary underlying assumption of these models, and the results indicate the 'average' component probabilities over the recording period. However, stationarity is required for several particular quantal models (e.g. simple and compound binomial models), and stationarity tests must be applied to the data before using these models. This can be done by calculating a regression fit to a plot of the sequentially measured synaptic potential amplitudes, or by calculating the average peak amplitude for successive groups of samples (e.g. 100 samples). If there is any significant trend in the amplitude during the recording period, or if the average amplitude of the subgroups differs significantly from the whole, then the data must be treated as non-stationary, and should not be analysed using the compound and simple bionomal models described below (21, 37, 43, 46). A further useful test is to compare the deconvolution results for successive samples during the recording period. In addition to this trend analysis, a test of randomness may also be useful (21, 43, 46).

Compound binomial quantal model
The generalized quantal model does not make assumptions concerning the probabilistic basis of the quantal amplitude fluctuations. A compound binomial model makes the further assumpion that the quantal fluctuations are due to the sum of statistically independent quanta chose from a *fixed* total population of available quanta, N (52). The probability, p_i, of occurrence of each population member (quantum), is independent of all others. Thus, the compound binomial distribution is described by a total population of available quanta, N, and the individual probabilities of all of the population members, $p_i = p_1, p_2 \ldots p_N$ (52).

For example, consider a model in which N represents the number of release sites, say 2, and that only a single quantum of transmitter can be released from each release site with probabilities of release p_1 and p_2, respectively. Then the probability of observing no quanta is $P_0 = (1-p_1)(1-p_2)$, the probability of observing the release of one quantum is $P_1 = p_1(1-p_2) + (1-p_1)p_2$, and the probability of observing the release of both quanta is $P_2 = p_1p_2$. In this case, the observed (measured) fluctuation pattern is represented by the compound binomial probabilities P_0, P_1, and P_2, with each component being a function of *all* of the underlying individual release probabilities $(p_1, p_2 \ldots p_N)$.

The noise-contaminated PDF for a compound binomial quantal model with a total population N is then:

$$f(V|\phi) = \sum_{i=0}^{N} P_i N(V-iQ)$$

where Q is the quantal amplitude, $N(V)$ is the noise PDF, and the P_i are given by:

$$P_0 = (1-p_1)(1-p_2) \ldots (1-p_N)$$
$$P_1 = p_1(1-p_2)(1-p_3) \ldots (1-p_N)$$
$$+ (1-p_1)p_2(1-p_3) \ldots (1-p_N)$$
$$+ \ldots$$
$$+ (1-p_1)(1-p_2)(1-p_3) \ldots p_N$$
$$P_{N-1} = p_1p_2 \ldots p_{N-1}(1-p_N)$$
$$+ \ldots$$
$$+ (1-p_1)p_2p_3 \ldots p_N$$
$$P_N = p_1p_2p_3 \ldots p_N$$

(see reference 52). Therefore, given the quantal size Q and the noise PDF, $N(V)$, the procedure of fitting a compound binomial distribution to the measured data is to determine the maximum likelihood estimates for N and all of the underlying quantal probabilities $p_1, p_2 \ldots p_N$. The application of the E–M algorithm to the compound binomial distribution can be found in reference 40.

Following (indirect) evidence at a connection in the spinal cord for a lack of significant quantal variability, but contaminating noise which was usually non-Gaussian, Walmsley *et al.* (20) developed and applied a maximum likelihood deconvolution procedure to examine the applicability of a compound binomial distribution (see *Figure 3*). An important general result from this study is that, while a compound binomial distribution may provide an excellent fit to the measured data, a range of acceptable solutions can usually be obtained with different values of N.

This problem has also been addressed by Smith *et al.* (43), who describe a maximum likelihood procedure which incorporates normally distributed contaminating noise and normally distributed quantal variance (assumed Type I, see Section 5.5). In agreement with Walmsley *et al.* (20), Smith *et al.* (43) found good fits to their data for different values of N, but have further suggested a penalty function for estimating the 'correct' value of N.

Careful consideration should be given to the interpretation of the 'correct' value of N. For example, if many release sites are completely silent, i.e. extremely small or zero release probability under the experimental conditions examined, then *at best*, the value of N determined using a compound binomial distribution equates with the number of release sites with release probabilities significantly greater than zero (referred to as the number of 'working' release sites by Walmsley *et al.* (8, 20). Thus, structural interpretation of the results should be made cautiously, and it would obviously be valuable to obtain independent anatomical information on the number of presumed release sites at the connection being studied.

Application of the compound binomial model to the analysis of fluctuations of synaptic potentials has generally been made to examine the proposal that the quantal probability is non-uniform among the population of available quanta (see references in 8, 43). It has been proposed that such differences in release probability may be related to structural features, such as differences in the area of release sites at the same connection (8). It should be noted that the possibility of non-uniform quantal release probabilities was originally raised by del Castillo and Katz (2) in their work on quantal release at the frog neuromuscular junction.

The compound binomial model described in this section represents the most general form. More specific compound binomial models have also been described which make further assumptions concerning the nature of the distribution of the underlying release probabilities (see Section 5.6).

Bruce Walmsley

Simple binomial quantal models

The simple or uniform binomial quantal model represents the most constrained quantal model. It is a specific case of the compound binomial model in which *all* of the underlying quantal release probabilities are assumed to be *identical*. In this case, the simple binomial distribution is defined by a (fixed) total population of available quanta, N, and a quantal probability, p, which is the same for each population member. The simple binomial probability distribution describes the probability of the ith multiquantal component as:

$$P_i = \frac{N!}{i!(N-i)!} p^i q^{(N-i)}$$

where $q = 1-p$ and $i = 0, 1, 2 \ldots N$ (see reference 52), and $f(V \mid \phi)$ is given by:

$$f(V \mid \phi) = \sum_{i=0}^{N} P_i N(V - iQ)$$

where Q is the quantal amplitude and $N(V)$ is the noise PDF.

Application of the simple binomial quantal model to quantal analysis is similar to the compound binomial model except that only a single release probability (rather than N release probabilities) needs to be estimated. The assumption of uniform quantal release probability represents a considerable constraint on this problem and this simplifies the fitting procedure (see reference 40 for application of the E–M algorithm).

The simple binomial model has found widespread use in the quantal analysis of synaptic potentials and currents and a detailed discussion of a variety of standard techniques for estimating p and N is provided by (1, 2, 5).

Korn and Faber (5) have described and used a maximum likelihood deconvolution procedure which assumes that quantal variability is negligible and which uses estimated noise (assumed to be normally distributed). Significantly, the anatomical details of each connection were examined, and the value of N obtained from the quantal analysis was well matched by the number of synaptic boutons contacting each cell (5). Korn and Faber (5) concluded that the value of N determined by the simple binomial analysis equates with the number of release sites (each bouton contains only a single release site at this connection), and that the release probability is identical at all release sites.

At a variety of other synaptic connections, application of the simple binomial model has resulted in the rejection of this model (*Figure 3*, see references in 8 and 43). Because the simple binomial is the simplest and therefore the most generally used quantal model, it is worth emphasizing that

Quantal analysis of synaptic transmission

a good fit of this model (or any model) to the data should not be interpreted as a *verification* of the initial assumptions (e.g. identical probabilities and an invariant population, N). Instead, the model should be viewed as being *compatible* (*or not*) with the assumptions, and independent evidence should be obtained for support (see 5).

5.5 Quantal variability and spontaneous potentials (currents)

A number of factors need to be considered when incorporating quantal variability into fluctuation analysis, even if quantal amplitude and variability have been determined directly from measurements of spontaneous synaptic potentials.

Variability in the measured amplitude of a quantal postsynaptic potential (current) is due to three main factors:

(a) Type I: variability between successive trials at the same release site
(b) Type II: differences in the mean amplitude (and variability) of the postsynaptic response between release sites
(c) contaminating noise (electronic and membrane noise)

It is important to recognize that the way in which quantal variability is incorporated into any quantal model depends on the origin of this variability.

Consider the effects of quantal variability in a noise-free situation. If the nature of the postsynaptic quantal response is exactly the same at all release sites (e.g. normally distributed with an identical mean amplitude, μ_Q, and variance, σ^2_Q), and if quantal responses at all release sites are statistically independent, then the variance of the nth multiquantal component in the evoked response is $n\sigma^2_Q$ (see *Figure 1*). If there is a total population of available quanta, N, then the variance of the maximum amplitude component is $N\sigma^2_Q$. However, if the variability in quantal amplitude is due to differences *between* release sites (Type II), then this calculation is not correct. Consider the example of a connection with N release sites. If *all* of the variability in the measured single quantal amplitudes is due to differences in the postsynaptic response between release sites (e.g. different numbers of activated receptors at different release sites, electrotonic effects, etc.), then the variance of the measured (spontaneous) quantal potential may actually be *greater* than that of the multiquantal peaks, and the variance of the maximum (N^{th}) multi-quantal component is zero.

In most cases there is also a significant contribution to the measured variability of the (spontaneous) quantal amplitude from contaminating noise. It may be possible to determine the noise-free spontaneous potential (current) PDF by application of one of the maximum likelihood deconvolution procedures (Section 5.1).

For example, if the amplitude distributions of the noise and the single quantal responses can both be described by normal distributions (assuming Type I variability), then the mean, μ_Q, and variance, σ^2_Q, of the quantal PDF are obtained. Subsequent analysis of the *evoked* potentials (currents) can then be achieved using the generalized quantal model (Section 5.4.2):

$$f(V \mid \phi) = \sum_{i=0}^{k} \frac{P_i}{\sqrt{2\pi\sigma^2_i}} e^{-(V-\mu_i)^2/2\sigma^2_i}$$

where $\sigma^2_i = \sigma^2_N + i\sigma^2_Q$, $\mu_i = i\mu_Q + \mu_N$, μ_Q and σ^2_Q are the mean and variance of the quantal PDF, μ_N and σ^2_N are the mean and variance of the noise PDF, and the unknown parameters are P_i, $i = 0, 1, 2, \ldots k$.

Further constraints on the P_i's are determined by the particular model chosen (generalized, compound binomial or simple binomial, see Section 5.4.2). See reference 40 for application of the E–M algorithm to this problem. Smith *et al.* (43) have described a quasi-Newton maximum likelihood procedure for a compound binomial quantal model, with normally distributed noise and quantal variability.

Serious errors in the subsequent fluctuation analysis will arise if the assumption of normal distributions is not correct, e.g. in some cases the amplitude distribution of spontaneous potentials is positively skewed. In this case, reliable measurements of the actual noise and/or the single quantal amplitude distributions should be used, or more appropriate explicit functions fitted to these distributions (40). Even if the noise contamination is reliably removed from the measurements of spontaneous synaptic potentials (currents), the origin of the remaining intrinsic quantal variability must be addressed (Type I or Type II), before it can be reliably incorporated into the subsequent quantal fluctuation analysis of evoked responses.

Figure 4 illustrates an example of incorporation of measured noise and

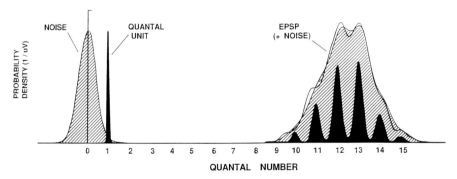

Figure 4. Incorporation of quantal variance (normally distributed Type I, see Section 5.5) and contaminating noise in a generalized quantal model. Solid black area, noise-free multiquantal PDF; shaded area, proposed noise-contaminated PDF; smooth curve, measured synaptic potential amplitude distribution (see Section 5.5 for further explanation.)

quantal variance (assumed to be normally distributed Type I) into the generalized quantal model described in the previous section (5.4.2).

5.6 Other models and procedures

The models described in Sections 5.4 and 5.5 represent those most commonly applied to the analysis of fluctuations in the amplitude of postsynaptic potentials and currents. Although one (or more) of these models may satisfactorily describe the observed fluctuations, they invariably represent a simplified description of synaptic transmission. A variety of other models have been suggested which incorporate non-uniform and/or non-stationary release properties.

Non-uniformities in quantal probabilities (p) are represented by the generalized compound binomial model (Section 5.4.2.*ii*), in which no *a priori* constraint is placed on the distribution of underlying probabilities. More specific compound binomial models have geen described which place further restriction on these probabilities. The simplest of these consists of the sum of two or more homogeneous (or simple binomial) subpopulations, each subpopulation having different average probability. Compound binomial distributions have been described in which p itself has a continuous distribution over the allowable range, 0–1 (52, see also references in (20, 43)). It is unlikely that the quantal probabilities are *absolutely identical* at any connection (although this may be approximately correct at some connections), and Miyamoto (21) has suggested that a normal distribution of release probabilities may reasonably represent the biological variation in p. In contrast, application of the generalized compound binomial model by Walmsley *et al.* (20) produced a highly skewed distribution of p values. However, in view of the uncertainties of such modelling (20), determination of the nature of the distribution of p to obtain specific compound binomial models will clearly require more direct measurements of the release characteristics at individual release sites (53, 54), in combination with a knowledge of the number of release sites, determined by ultrastructural analysis.

Under a variety of physiological and experimentally induced conditions, the nature of release varies with time, e.g. variation in release probabilities or the population of 'available' quanta, or variation in the amplitude of the quantal postsynaptic potential (current). While a lack of stationarity may be obvious from trends in the measurements of synaptic potentials (currents, see Section 5.4.2), moment-to-moment fluctuations in, say, N and/or p may not be detected by this analysis. Several studies have examined the effects of non-stationarity, and a detailed discussion of this important issue can be found (2, 47, 48, 55, 56). In general, these studies emphasize that a variety of quantal models incorporating non-uniformity of release probabilities and/or non-stationarity of release probabilities and N may be fitted to a given set of data.

One of the main aims of quantal analysis is to reveal changes in the amplitude of quantal postsynaptic potentials (currents) and/or in the quantal probabilities, following changes at a connection (e.g. facilitation. LTP, drug action). The generalized quantal model (Section 5.4.2) does not assume stationarity in p or N, and may be useful in indicating a lack of change in quantal amplitude, and/or a change (on average) in the probabilistic nature of the fluctuations (including an estimation of the proportion of failures, if they occur). However, considerable caution must be exercised in this analysis, particularly under the conditions of poor, and/or changing signal-to-noise ratio conditions. An apparent lack of change of the average quantal size may be due to a resolution artefact of the method, and/or problems associated with small sample size. If the amplitude of the quantal postsynaptic potentials has changed, careful consideration must be given before applying a model which assumes that the quantal amplitude has changed to the *same* new average value at *all* release sites. Large changes in the quantal amplitude at some release sites and a small or no change at others will obviously compound this problem (see Sections 5.4.2 and 5.5).

Finally, as outlined in Section 3, the interpretation of a change in quantal size or probability as being pre- or postsynaptic in origin is entirely model dependent. A detailed knowledge of the ultrastructural details of a connection is obviously an advantage in the interpretation of this issue and in the use of models of synaptic transmission in general.

Acknowledgements

I would like to thank Drs R. J. Sayer and D. M. Kullmann for their helpful comments on this manuscript, and the National Health and Medical Research Council of Australia for financial support.

References

1. Martin, A. R. (1977). In *The Handbook of Physiology, Sect. 1. The Nervous System*, Vol. 1 (part 1) (ed. E. R. Kandel), pp. 329–355. American Physiological Society, Bethesda.
2. McLachlan, E. M. (1978). In *International Review of Physiology, Neurophysiology III*, Vol. 17 (ed. R. Porter), pp. 49–117. University Park Press, Baltimore.
3. Silinsky, E. M. (1985). *Pharmacol. Rev.*, **37**, 81.
4. Atwood, H. L. and Wojtowicz, J. M. (1986). In *International Review of Neurobiology*, Vol. 28 (ed. J. R. Smythies and R. J. Bradley), pp. 275–362. Academic Press, New York.
5. Korn, H. and Faber, D. S. (1987). In *Synaptic Function* (ed. G. Edelman, E. Gall, and W. M. Cowan), pp. 57–108. Wiley, New York.
6. Stjärne, L. (1989). *Rev. Physiol. Biochem. Pharmacol.*, **112**, 1.
7. Redman, S. J. (1990). *Physiol. Rev.*, **70**, 165.

8. Walmsley, B. (1991). *Prog. Neurobiol.*, **36**, 391.
9. Jack, J. J. B., Kullmann, D. M., Larkman, A. U., Major, G., and Stratford, K. J. (1990). *Cold Spring Harbor Symposium on Quantitative Biology*, Vol. 55 pp. 57–67. Cold Spring Harbor Laboratory Press, Cold Spring Harbour.
10. Landis, D. M. D., Hall, A. K., Weinstein, L. A., and Reese, T. S. (1988). *Neuron*, **1**, 201.
11. Hirokawa, N., Sobue, K., Kanda, K., Harada, A., and Yorifuji, H. (1989). *J. Cell Biol.*, **108**, 111.
12. Tatsuoka, H. and Reese, T. S. (1989). *J. Comp. Neurol.*, **290**, 343.
13. Luff, S. E., McLachlan, E. M., and Hirst, G. D. S. (1987). *J. Comp. Neurol.*, **257**, 578.
14. Seitanidou, T., Triller, A., and Korn, H. (1988). *J. Neurosci.*, **8**, 4319.
15. Land, B. R., Salpeter, E. E., and Salpeter, M. M. (1981). *Proc. Natl. Acad. Sci. USA*, **78**, 7200.
16. Edwards, F. R., Redman, S. J., and Walmsley, B. (1976). *J. Physiol.*, **259**, 665.
17. Edwards, F. R., Redman, S. J., and Walmsley, B. (1976). *J. Physiol.*, **259**, 689.
18. Edwards, F. R., Konnerth, A., and Sakmann, B. (1990). *J. Physiol.*, **430**, 213.
19. Matteson, D. R., Kriebel, M. E., and Llados, F. (1981). *J. Theor. Biol.*, **90**, 337.
20. Walmsley, B., Edwards, F. R., and Tracey, D. J. (1988). *J. Neurophysiol.*, **60**, 889.
21. Miyamoto, M. (1986). *J. Theor. Biol.*, **123**, 289.
22. Brown, A. G. and Fyffe, R. E. W. (1984). *Intracellular Staining of Mammalian Neurones*. Academic Press, London.
23. Redman, S. J. and Walmsley, B. (1981). *Trends Neurosci.*, **4**, 248.
24. Redman, S. J. and Walmsley, B. (1983). *J. Physiol.*, **343**, 117.
25. Redman, S. J. and Walmsley, B. (1983). *J. Physiol.*, **343**, 135.
26. Purves, R. D. (1981). *Microelectrode Methods for Intracellular Recording and Ionophoresis*. Academic Press, London.
27. Smith, T. G. (Jnr), Lecar, H., Redman, S. J., and Gage, P. W. (ed.) (1985). *Voltage and Patch Clamping with Microelectrodes*. American Physiological Society, Bethesda.
28. Solodkin, M., Jiminez, I., Collins, W. F. III, Mendell, L. M., and Rudomin, P. (1991). *J. Neurophysiol.*, **65**, 927.
29. Edwards, F. R., Hirst, G. D. S., and Silinsky, E. M. (1976). *J. Physiol.*, **259**, 647.
30. Hackett, J. T., Cochran, S. L., and Greenfield, L. J. (1989). *Neuroscience*, **32**, 49.
31. Barrett, E. F. and Stevens, C. F. (1972). *J. Physiol.*, **227**, 665.
32. Cope, T. C. and Mendell, L. M. (1982). *J. Neurophysiol.*, **47**, 455.
33. Walmsley, B. and Stuklis, R. (1989). *J. Neurophysiol.*, **61**, 681.
34. Koch, C. and Segev, I. (ed.) (1989). *Methods in Neuronal Modeling*. MIT Press, Cambridge.
35. Everitt, B. S. (1987). *Introduction to Optimization Methods and their Application in Statistics*. Chapman and Hall, London.
36. McLachlan, G. J. and Basford, K. (1987). *Mixture Models: Inference and Applications to Clustering*. Marcel Dekker, New York.
37. Kendall, M. and Stuart, A. (1977). *The Advanced Theory of Statistics*. Charles Griffin, London.

38. Titterington, D. M., Smith, A. F. M., and Makov, U. E. (1985). *Statistical Analysis of Finite Mixture Distributions*. John Wiley, Chichester.
39. Ling, L,. and Tolhurst, D. J. (1983). *J. Neurosci. Methods*, **8**, 309.
40. Kullman, D. M. (1989). *J. Neurosci. Methods*, **30**, 231.
41. Bunday, B. D. (1984). *Basic Optimization Methods*. Edward Arnold, London.
42. Fletcher, R. (1987). *Practical Methods of Optimization*. John Wiley, Chichester.
43. Smith, B. R., Wojtowicz, J. M., and Atwood, H. L. (1991). *J. Theor. Biol.*, **150**, 457.
44. Clamann, H. P., Rioult-Pedotti, M.-S., and Luscher, H.-R. (1991). *J. Neurophysiol.*, **65**, 67.
45. Press, W. H., Flannery, B. P., Teukolsky, S. A., and Vetterling, W. T. (1989). *Numerical Recipes*. Cambridge University Press.
46. Sugimoto, H., Ishii, N., Iwata, A., and Suzumura, N. (1977). *Computer Programs Biomed.*, **7**, 293.
47. Barton, S. B. and Cohen, I. S. (1977). *Nature*, **268**, 267.
48. Brown, T. H., Perkel, D. H., and Feldman, M. W. (1976). *Proc. Natl. Acad. Sci. USA*, **73**, 2913.
49. Walmsley, B., Edwards, F. R., and Tracey, D. J. (1987). *J. Neurosci.*, **7**, 1037.
50. Clements, J. D. (1990). *J. Neurosci. Methods*, **31**, 75.
51. Van der Kloot, W. (1989). *J. Neurosci. Methods*, **27**, 81.
52. Johnson, N. L. and Kotz, S. (1969). *Discrete Distributions*. Houghton Mifflin, Boston.
53. Bennett, M. R. and Lavidis, N. A. (1982). *Dev. Brain Res.*, **5**, 1.
54. D'Alonzo, A. J. and Grinnell, A. D. (1985). *J. Physiol.*, **359**, 235.
55. Perkel, D. H. and Feldman, M. W. (1979). *J. Math. Biol.*, **7**, 31.
56. Higashima, M., Sawada, S., and Yamamoto, C. (1990). *Neurosci. Lett.*, **115**, 231.

7

Intracellular staining of enteric neurones

GORDON M. LEES

1. Introduction

It is now widely accepted that the enteric nervous system (ENS) provides neuroscientists with many challenges, one of the most important problems being the identification of neuronal pathways involved in intestinal reflexes, particularly intrinsic reflex responses, such as peristalsis. The intrinsic neurones of the gastrointestinal tract lie in ganglionated plexuses between the (outer) longitudinal muscle and the circular muscle layers (myenteric plexus) and in the submucosa (inner and outer submucous plexuses) (*Figure 1*). Morphologically, electrophysiologically, immunohistochemically, and pharmacologically these enteric neurones show an amazing diversity, both between and within the plexuses (1–10).

Recent investigations of the cross-correlations between these properties have significantly advanced current knowledge of the complex organization of the ENS (11–26). These have been achieved mainly by the application of intracellular staining and intracellular electrophysiological recording techniques, combined with immunohistochemical (IHC) methods developed for whole-mount preparations. This has led to a better understanding of the properties of particular types of enteric neurones, including their synaptic connections with other neurones. In this context, it must be appreciated that, unlike the neurones of the central nervous system (CNS), clusters of morphologically, electrophysiologically, or neurochemically similar enteric neurones can seldom be identified in tissue that has not been specially processed.

2. Equipment required

- dissection microscope (magnification 3–50) with incident and transmitted lighting
- a small, shallow, organ bath lined with Sylgard® (Dow Corning) and at least one platinum wire electrode partially exposed (*Figure 2*)

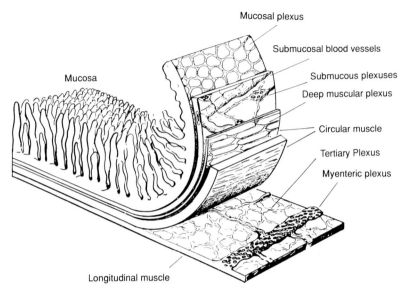

Figure 1. Schematic representation of guinea-pig small intestine separated into layers to show the principal features (for further details, consult references 8 and 19). Note that the cell bodies of enteric neurones are confined to the myenteric plexus and submucous plexuses.

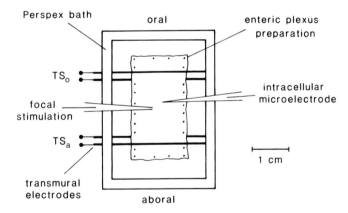

Figure 2. Diagram of small organ bath used for intracellular recording and staining of enteric neurones. The base has a rectangular hole into which is sealed a microscope coverslip, through which transmitted light can pass. The bath is lined with Sylgard, to which the preparation is pinned. The platinum wire electrodes for transmural stimulation oral (TS_o) and anal (TS_a) to the point of recording pass through holes in the walls of the bath where they are sealed in place to prevent leaks. The lower electrodes are exposed only in the middle of the bath, by cutting a slit in the Sylgard lining. The upper electrodes, which should be hemi-insulated, are held in gentle contact with the preparation and directly above the lower ones by means of Λ-shaped pins.

- a compound microscope (preferably fitted with a high-pressure mercury lamp for epi-fluorescence with suitable filters (*Table 1*) and a suitable objective for viewing individual ganglia at a magnification of 200 to 400); a calibrated eyepiece graticule would be advantageous
- a micromanipulator with X, Y, Z movements
- a microelectrode puller capable of producing ultra-fine microelectrodes from 1.0 mm (o.d.) glass capillary tubing
- a microelectrode preamplifier with current injection facility and extensive capacitance neutralization
- an oscilloscope with a timebase range of 5 msec/div to 2 sec/div
- a two-channel physiological stimulator capable of providing DC current and pulses of up to at least 200 msec duration
- a pen recorder or plotter, preferably also a tape-recorder or modified video-cassette recorder
- a heater–circulator connected to a heat-exchange coil (20–50 ml)

Table 1. Suggested Zeiss filter combinations for fluorescent intracellular markers[a]

Fluorophore	Excitation	Dichroic mirror	Emission	Comments
Lucifer Yellow CH	LP418 + KP490 (or filter set 05)	FT510	LP520 ± KP560	Use BG12 (3–6 mm thick) or ND (0.05–0.2)
Carboxyfluorescein	LP418 + KP490	FT510	LP520	Use BG12 (3–6 mm thick) or ND (0.05–0.2)
Biocytin[b]	BP530 − 585 (or filter set 00)	FT600	LP615	

[a] ND, neutral density filter; LP, long pass (cut-on); KP, short pass (cut-off); the combination of an LP and a KP filter effectively creates a band pass filter between the stated wavelengths (nm); BP, interference band pass filter, which transmits light between stated wavelengths (nm); FT, reflects below and transmits (passes) above stated wavelength (nm). BG 12 is a dyed glass filter available in thicknesses of 1–3 mm, thus two should be used together.
[b] Used with streptavidin coupled to Texas red.

3. Solutions, materials, and dyes

3.1 Krebs' solution

Isolated tissues should be bathed or continuously superfused at 3–10 ml/min with Krebs' solution (35–37 °C) of the following composition (mM): NaCl, 118; KCl, 4.75; $CaCl_2$, 2.54; $MgSO_4$, 1.2; NaH_2PO_4, 1.0; $NaHCO_3$, 25.0; glucose, 11.1. The solution should be bubbled with 95% O_2/5% CO_2 (pH 7.36–7.40). For myenteric plexus preparations, the Krebs' solution should also contain nicardipine (0.5 μM), hyoscine (1 μM), and mepyramine (1 μM).

3.2 Other solutions required

- a modified Krebs' solution consisting of 143 mM NaCl; 4.75 mM KCl, and 2.54 mM $CaCl_2$
- 165 mM NaCl
- 0.5 M KCl
- Lucifer Yellow CH 2% (w/v; aqueous) and 0.5% (w/v; dissolved in 0.5 M KCl). The Lucifer Yellow solutions should be stored in Eppendorf (or similar) tubes and protected from light; these tubes can be centrifuged to remove the precipitate. Filtration is very wasteful of dyes

3.3 Other materials

- several dozen pins (2–4 mm long) will be required; cut these from tungsten wire (100, 50, and 25 μm diameter; Goodfellow Metals)
- Sylgard slabs (6 cm long, 3–4 cm wide, and 3–4 mm thick) should also be prepared for fixation of specimens

4. Procedures

4.1 Preparation of myenteric plexus

Protocol 1 describes the initial dissection and *Protocol 2* the arrangement of the preparation in the organ bath and the final stages in exposing the myenteric plexus. To reduce the difficulty in dissecting the mucosa and submucosa and to maximize the sensitivity of the preparation to acetylcholine, it is advantageous to use small guinea-pigs (180–250 g) that have been fed fresh vegetables, in addition to pellets, and have had access to drinking water with ascorbic acid supplements.

Protocol 1. Initial dissection and preparation

1. Noting the orientation, remove a segment (at least 5 cm long) of the gastrointestinal tract from an animal which has just been killed. Keep the issue immersed in Krebs' solution (32–37 °C) to minimize neuronal swelling and dysfunction.
2. Flush out the luminal contents with Krebs' solution (35–37 °C) *with minimal distension* and transfer the segment to a clean Petri dish; do not leave the gut in contact with the luminal contents. Ideally, the vasculature of the tissue (e.g. duodenum) should be perfused to flush out erythrocytes, which are an important source of autofluorescence and of degradative enzymes.
3. Pin the oral and anal ends of the segment to the base of the Petri dish or

the small organ bath lined with Sylgard (*Figure 2*) in such a way that the mesenteric border lies uppermost and in a straight line.

4. Cut along this border with straight-bladed (strabismus) scissors and pin the preparation flat with light longitudinal and radial stretch.
5. View the preparation, which should now be rectangular, at a magnification of 15–25 in transmitted light.
6. Using two pairs of *blunt*-ended jewellers' forceps, pick up the mucosa, together with the submucosa, close to the anal end and to one of the mesenteric borders. Gently peel the tissue back sufficiently to reveal the underlying circular muscle layer.
7. Repeat the procedure along the entire length of the specimen, avoiding the centre of the preparation. Take care not to damage the muscle layers; the smooth muscle will contract (appears darker) if stretched excessively, especially in the absence of the calcium-antagonist, nicardipine.
8. Next, peel the tissue as far as the mesenteric borders and gently ease the submucosa off the circular muscle at the edges of the preparation, repinning if necessary.

Protocol 2. Arrangement in the organ bath and final preparation

1. Arrange the preparation in the organ bath so that at least the oral end is above one of the transmural electrodes.
2. For the small intestine, go to step 3, for the colon go to step 5. It cannot be over-emphasized that the correct intensity and angle of lighting is critical for the success of steps 3–5.
3. Remove the *circular* muscle layer, over as large an area as is required for the intended experiment. Use high magnification and square-ended (i.e. not sharp-pointed), but very fine (No. 5) jewellers' forceps (Dumont Biologie). Commence at one mesenteric border, where the circular muscle may already be partially detached.
4. Peel away the muscle borders across the entire width of the preparation, at an angle of 10–15° (for best results). The exposed myenteric plexus will remain adherent to the longitudinal muscle. Proceed to step 6.
5. For the colon, it may be better to remove the *longitudinal* muscle because the myenteric plexus is often more firmly attached to the circular muscle layer.
6. Pin the upper transmural electrode(s) in place directly above the corresponding electrode beneath the tissue (see *Figure 2*).
7. Seal the bath before transferring it to the stage of the compound microscope and commencing the superfusion.

4.2 Preparation of submucous plexuses

The preparation of the submucous plexuses is essentially the same as for the myenteric plexus, except that the mucosa is best dissected off the submucosa in longitudinal strips using a single pair of forceps. Try to clear an area about 10–18 mm long and 8–12 mm wide, then turn the preparation over to dissect away the longitudinal and circular muscle layers. Finally, free the submucous preparation of any adherent strands of circular muscle; re-pin, if necessary, mucosal side downwards.

It should be noted that, although ganglionated plexuses have been reported to exist at different levels within the submucosa of larger mammals (27–33), you may not be able to distinguish these in whole-mount preparations of submucosa from guinea-pig ileum for morphological and electrophysiological studies. Submucous plexus preparations of the duodenum pose many problems in dissection, recording, fixation, and immunohistochemistry, and are, therefore, not recommended.

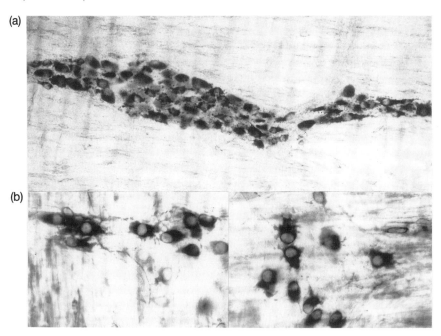

Figure 3. Examples of guinea-pig myenteric plexus neurones revealed by the diaphorase technique (nitro-blue tetrazolium). (a) Ileum. Note how well defined the ganglion is due to dense packing of neurones. A high proportion show the colour reaction, with the typical preponderance of Dogiel Type II cells (large, smooth, ovoid soma). (b) Two preparations of the descending colon between taeniae. Note the looser association of neurones forming the ganglia and the higher incidence of Type I cells (numerous, short, lamellar processes). (Figure courtesy of R. J. May and Shauna M. C. Cunningham (unpublished observations).)

4.3 Histochemical detection of myenteric and submucous plexuses

A useful means of becoming familiar with the general appearance and disposition of enteric neurones is to visualize them using the nitro-blue tetrazolium (diaphorase) method (see *Figure 3*). The following modification (*Protocol 3*) of Gabella's original method (34) has proved to be more reliable and effective (May, Charleson, and Lees, unpublished observations).

Protocol 3. Visualization of neurones with the diaphorase method

Materials

- 9% sucrose solution: w/v in phosphate-buffered saline (PBS), pH 7.1
- NBT–NADH solution: 2.5 mg nitro-blue tetrazolium (NBT) in 5.0 ml distilled water, mixed immediately before use with 10 mg reduced nicotinamide dinucleotide (NADH) in 5.0 ml 0.1 M phosphate buffer (pH 6.4 or pH 7.3)
- neutral-buffered formalin
- buffered glycerol (pH 8.6)

Method

1. Expose either plexus preparations (by immersion) or whole-thickness specimens (by perfusion via mesenteric vasculature) to the 9% sucrose solution for 5–10 min.
2. Treat in the same way with the solution of NBT–NADH for 30–45 min[a].
3. Leave the preparations in neutral-buffered formalin for 12–18 h at 4 °C to stop the reaction.
4. Dissect further, if necessary. Then, mount the preparations in buffered glycerol.
5. View slides at a relatively high magnification (25–40 ×) to check quality of reaction.

[a] This procedure induces a selective colour reaction resulting from the action of NADH diaphorase on NBT. The enzyme activity is much greater in neurone somata, which therefore react earlier than other neural elements or other cells (e.g. smooth muscle).

Naturally, it is best to use fresh tissue for each experiment involving intracellular staining and recording techniques. In view of the high costs of certain animals, however, it would be advantageous if tissue could be stored in good condition for intracellular staining or recording for 24–72 h.

Intracellular staining of enteric neurones

Although the guinea-pig ileum has been stored successfully under organotypic culture conditions (35), thicker specimens would probably have to be perfused with washed, oxygenated erythrocytes in a saline solution (if necessary, containing dextran), in addition to being maintained in a culture medium, to minimize anoxic and osmotic damage.

4.4 Fixation and processing of specimens

If the stained preparation is not to be photographed at once, it should be fixed overnight in neutral-buffered formalin. The latter can be removed by washing in PBS (pH 7.1), prior to mounting in buffered glycerol (pH 8.6). Alternatively, if immunohistochemical studies are to be undertaken, the preferred fixative (e.g. Zamboni's) should be used; subsequently, alcohol and xylene dehydration procedures may be required, depending on the antibody used. For Zamboni's fixation, either *Protocol 4* or *Protocol 5* is known to work satisfactorily with several antibodies to neuropeptides (see also reference 37). Over-fixation and inadequate clearing of the fixative leads to an excessively high background fluorescence (autofluorescence), which may make satisfactory photomicrography impossible.

Protocol 4. Long Zamboni's fixation

Materials

Zamboni's fixative: formaldehyde (50 ml of 'Analar' grade solution available commercially as a 40% aqueous solution); 0.2 M sodium phosphate buffer, pH 7.0 (500 ml made from 0.2 M NaH_2PO_4 (195 ml) plus 0.2 M Na_2HPO_4 (305 ml); 150 ml saturated aqueous solution of picric acid (filtered); distilled water to 1 litre

Method

1. Pin out specimen on slab of Sylgard and fix in cold Zamboni's solution for 16–24 at 4 °C.
2. Wash and dehydrate for 45–60 min in 80% ethanol (it is essential to remove picric acid)[a].
3. Further dehydration and rehydration following the scheme: 30 min each in ethanol (95%, 100%, 100%), then xylene (twice); then 15 min each in ethanol (100%, 95%, 80%, 50%), then purified (or double-distilled) water, finally PBS[b].

[a] Zamboni's fixative can also be removed by soaking in dimethylsulphoxide (DMSO) (10 min, three times; reference 34).
[b] This is not a suitable medium for long-term storage of specimens; instead, use PBS containing 0.1% sodium azide, if it is necessary to keep specimens for several weeks prior to carrying out immunohistochemical procedures.

Protocol 5. Short Zamboni's fixation

This method may not be suitable for all antibodies.

1. Pin out specimen on slab of Sylgard and fix in cold Zamboni's solution for 4 h at 4 °C.
2. Wash and dehydrate for 45–60 min in 80% ethanol[a].
3. Further dehydration and rehydration using the following scheme: 15 min each in ethanol (95%, 100%, 100%), then xylene (twice); then 5 min each in ethanol (100%, 95%, 80%, 50%), then purified (or double-distilled) water, finally PBS[b].

[a] Zamboni's fixative can also be removed by soaking in DMSO (10 min, three times; reference 34).
[b] This is not a suitable medium for long-term storage of specimens; instead, use PBS containing 0.1% sodium azide, if it is necessary to keep specimens for several weeks prior to carrying out immunohistochemical procedures.

5. Ancillary techniques

5.1 Fluorescence microscopy

For the intracellular injection of fluorescent dyes, a suitable objective should have a high numerical aperture, relative to its magnifying power (e.g. × 25/0.6, or even better × 25/0.8) but it need not be of the high quality required for immunofluorescence. In the latter case, there is no substitute for special objectives for fluorescence microscopy of fixed specimens, for which immersion objectives (× 16, × 25, × 50) are very suitable. It is also advisable to have available neutral density filters to avoid unnecessary bleaching of the fluorophores and to reduce flare.

It is important to appreciate that, in some preparations, a particular fluorophore may be better viewed with certain filter combinations than with others. Lucifer Yellow is a good example of such a dye, which in low concentrations may be difficult to distinguish from tissue autofluorescence. It is useful to have a 'minus red' barrier filter (e.g. Zeiss KP560) for viewing this dye, especially when used in conjunction with rhodamine- or Texas red-coupled (labelled) secondary antibodies.

5.2 Fluorescence immunohistochemistry

Since details of this technique are beyond the scope of this chapter, the reader is advised to consult reference 38.

The fading of immunofluorescence can be reduced by the inclusion of special agents in the mounting medium, which is usually buffered glycerol (pH 8.6).

5.3 Photomicrography

Although stained cells may seem very brightly fluorescent and, therefore, easily photographed, initial attempts at photomicrography may be rather disappointing unless the following points are fully appreciated (see also reference 39).

(a) So-called 'fast' films, intended for daylight use, may perform poorly and may seem very much slower than expected, due to reciprocity failure. The most successful films are Kodak Plus-X, Ilford HP5, and FP4; Kodak T-Max 3200 should be rated at only 800, but the contrast and resolution are disappointing.

(b) The colour balance and sensitivity is likely to be markedly altered, so that one film may be ideal for blues and greens but not for reds. For superb results, try Agfachrome 1000RS, Agfapan 400 ISO, or *gas hypersensitized* Kodak Technical Pan 2415 (Lumicon), exposed at 24–28 DIN (200–640 ISO) and developed in Kodak D19 developer at 21 °C for 4–4.5 min. The Kodak 2415 film is relatively red sensitive.

6. Intracellular staining

6.1 Choice of intracellular markers

It is important to appreciate that the morphology of all neuronal types is not always reliably or satisfactorily visualized by histochemical or immunohistochemical procedures (*Figures 3* and *4*). Indeed, the intracellular distribution of many neurochemicals precludes an adequate revelation of the soma shape or size. Although many agents have been tried, few have proved to be either adequate for revealing neuronal morphology or suitable for correlative studies with intracellular electrophysiological techniques; fewer still are compatible with immunohistochemical methods. The attributes of the most successful dyes currently used for enteric neurone morphology are summarized, in the following sections.

6.6.1 Lucifer Yellow CH (see references 11, 40–42)

- a fluorescent dye with a high quantum yield
- fades noticeably in fixed tissue
- limited compatibility with potassium salts, hence tendency for electrodes to block
- good morphology, with little evidence of spillage if the dye is applied extracellularly
- usually good electrophysiology when used in low concentrations and dissolved in weak solutions of KCl

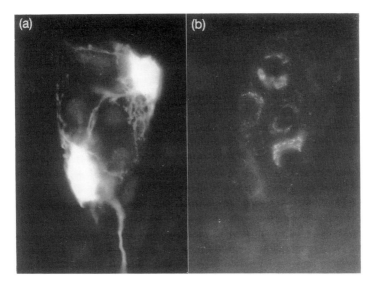

Figure 4. Comparison of the morphological information obtainable from immunohistochemical studies (see also *Figure 6*). Although the morphology of the neurones illustrated was not known, previous extensive studies with Lucifer Yellow staining have shown that neurones showing immunoreactivity to these neuropeptides are morphologically indistinguishable (13, 61, 62). (a) Submucous plexus neurones of guinea-pig ileum showing immunoreactivity to neuropeptide Y (NPY). Note the remarkable detail of the short soma processes. Polyclonal anti-NPY antibody (raised in rabbit), 1:1000; secondary antibody, goat anti-rabbit IgG labelled with fluorescein isothiocyanate (FITC) as fluorophore. (b) Identical field and focus, showing neighbouring neurones with immunoreactivity to vasoactive intestinal peptide (VIP). Note the presence in the cytoplasm of immunoreactivity only in the immediate vicinity of the nucleus. Monoclonal anti-VIP antibody (raised in mouse ascites cells) VIP 31, 1:600; secondary antibody, goat anti-mouse IgG labelled with rhodamine as fluorophore. (Figure courtesy of P. J. Campbell and G. M. Lees (unpublished observations).)

- compatible with fluorescence IHC but, with prolonged exposure to exciting blue light, tends to develop prominent emission at long wave-lengths (*Figure 5*), thus sometimes producing a 'false positive' image in IHC
- antibodies available to convert dye to dark reaction product for permanent visualization of morphology of excellent quality

6.1.2 5,6-Carboxyfluorescein (Dr S. J. H. Brookes, personal communication)

- a fluorescent dye with high quantum yield
- does not fade rapidly
- compatible with potassium salts, hence gives excellent electrophysiological recordings

Intracellular staining of enteric neurones

- very good morphology
- excitation and emission spectra often too broad to permit use with fluorescence IHC (G. T. Pearson, D. J. Leishman, and G. M. Lees, unpublished observations)

6.1.3 Biocytin (26)

- not a dye, but can be the substrate for an IHC reaction that reveals the morphology as either a fluorescent or permanent reaction product
- most useful when combined with a low concentration (0.2–0.5%) of either Lucifer Yellow or a mixture of carboxyfluorescein (1–2%) and KCl (0.5–2 M)
- compatible with potassium salts, hence gives excellent electrophysiological recordings
- superlative revelation of morphology
- compatible with fluorescence IHC

6.1.4 Procion Yellow M4RX (S)

- fluorescent dye with a low quantum yield, previously used, which fades little
- forms gel with potassium salts in microelectrodes, hence noisy electrodes which tend to block and give poor quality electrophysiological records
- readily forms covalent bonds with extracellular, as well as intracellular, proteins to cause persistent staining, hence 'spillage' often obscures morphological detail
- little or no loss from damaged cells
- excitation and emission spectra too broad to be suitable for combination with fluorescence

6.1.5 Horseradish peroxidase (12)

- non-fluorescent dye made to form a reaction product detectable in light or electron microscope
- not easy to fill ultra-fine microelectrodes required for enteric neurones
- requires large, depolarizing current pulses or pressure ejection system
- often gives poor 'staining' of enteric neurones
- compatibility with IHC limited, none with fluorescence IHC

6.2 Requirements and precautions

Generally, microelectrodes for use with the above intracellular markers are pulled from borosilicate glass (1.0 mm o.d., 0.5–0.7 mm i.d.) and have a higher resistance than those filled with 0.5–2.0 M KCl (40–150 MΩ), alone; hence, they are noisier, need greater capacitance neutralization and tend not to pass both depolarizing and hyperpolarizing current equally well. Typically, with Lucifer Yellow or carboxyfluorescein, currents of 0.02–0.5 nA (passed for 0.2–0.8 sec at 1 Hz for up to 30 sec) are required to produce an adequate stain. Alternatively, 25–35 kPa may be needed in pressure ejection systems.

Microelectrodes are filled by injecting a tiny volume from a 1 ml syringe (solution protected from light), fitted with a gas chromatography needle (Hamilton, KF 730, point style No. 3, 51 mm long) or a long, fine hypodermic needle until the solution fills the shank as far as the shoulder of the microelectrode. The rest of the electrode is back-filled with an electrolyte solution. In the case of Lucifer Yellow, persons with no previous experience in intracellular staining are recommended to use a 1–3% aqueous solution, back-filled with 0.165 M NaCl, rather than LiCl; such electrodes do not block as readily, pass current easily, and produce excellent staining (*Figure 5*). For electrophysiological recording, however, it is advisable to use Lucifer Yellow dissolved in KCl (see Section 3). Fill the microelectrode with the dye solution free from precipitate.

6.2.1 Exposure to blue light

When using Lucifer Yellow and other fluorescent dyes do not expose unfixed tissue to ultraviolet light or to light-containing wave-lengths at the violet–blue end of the visible spectrum (400–490 nm), particularly if focused by a condenser lens, for more than just a few seconds *in toto* to avoid:

- bleaching of the dye in stained cells and in the tip of the microelectrode
- damage to cells already injected with the fluorophore because fluorescence causes local heating, which can kill the cell (43)
- damage to unstained, neighbouring neurones and glia

6.2.2 Current limitations

Do not use more current than is absolutely necessary to eject the dye into a cell because this will lead to:

- decomposition of the dye in the micrelectrode tip; in the case of Lucifer Yellow, dye in the tip will fluoresce orange–red
- over-filling of the cell, the damage being manifest as either swelling of the soma or of processes (bloated appearance), especially of varicosities or vacuolation of the cell

Staining of the nucleus is very common with Lucifer Yellow and carboxyfluorescein and, in the case of Lucifer Yellow, this can often be advantageous for IHC (see below and *Figures 5* and *6*).

6.3 Quality control

Note carefully the point of impalement of a neurone (soma or soma process close to, or far from, the soma) and whether more than one cell has been stained during the attempt to impale a single neurone. It is not uncommon for a neurone to appear to change its electrophysiological properties, only for the investigator to discover that the microelectrode has slipped out of one neurone into another (see end of Section 7.3). This must not be confused with dye-coupling of cells, which is commonly seen in glial cells.

Try varying the angle or depth of impalement in myenteric ganglia; superficial impalements are likely to be of glial cells or Dogiel Type II cells, whereas a steeper angle of attack or deeper impalement is more likely to result in the staining of neurones of other types (see Section 6.5).

As indicators of the efficiency of your staining, measure the length of the long soma processes and compare (a) the detailed morphology of the short soma processes, (b) the fineness of processes between varicosities or lamellar expansions of processes, (c) the sharpness of outline of varicosities, and (d) the fineness of branching that can be detected, with those of references 11–19, 21, 24.

Even when experienced, anticipate unsatisfactory results, such as:

(a) poor staining:
   i. faint soma, even fainter on withdrawal of microelectrode
   ii. re-impalement may or may not help, depending on the degree of prior injury
   iii. no clear filling of soma processes
(b) damage:
   i. swollen appearance of soma or varicosities near soma
   ii. spillage around soma or point of impalement (often in the form of fine speckling)
   iii. vacuoles (non-fluorescent regions) in soma

6.4 Procedures for intracellular staining

Do not attempt to impale cells under direct visual control. They are likely to be extensively damaged. It is best to proceed as if primarily intending to record electrophysiologically (see Section 7.1)

6.5 Cell types most likely to be stained

Neurones, glial cells, and smooth muscle cells are most likely to be stained (see *Figure 5*). Occasionally, very thin, rectangular cells may be impaled; these are thought to be fibroblasts. Special techniques are required to impale interstitial cells of Cajal.

6.6 Problems

Inadequate staining/filling is most commonly due to insufficient current actually passed (the tip may be partially blocked or the dye may have been

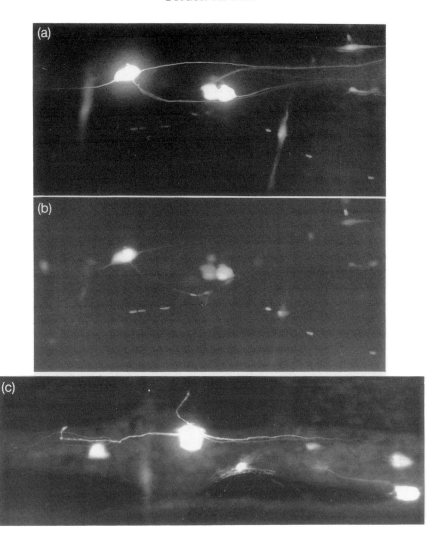

Figure 5. Examples of cell types likely to be stained in myenteric plexus preparations. (a) The three brightly stained cells are Dogiel Type II neurones with long soma processes running circumferentially (anal at top). Other cell types stained (left to right): longitudinal muscle fibre (out of plane of focus), longitudinal muscle fibre, circular muscle fibre with faint dye coupling (above), and a glial cell (beneath). Lucifer Yellow (2% w/v) staining, microelectrode back-filled with 0.165 M NaCl solution. Blue incident light. (b) The same field but viewed in green exciting light (Zeiss filter set 15 for rhodamine). Note that all seven cells can be seen, the nuclei being clearly visible in the neurones. For significance, see text Section 6.1. (c) Lucifer Yellow-stained cells in another guinea-pig myenteric plexus preparation of ileum. Note the comparative sizes of the somata and soma processes of neurones and glial cell (lower centre) and, secondly, the typical branching pattern of the Dogiel Type II neurone (upper centre). (Figure courtesy of Kirsteen N. Browning (unpublished observations).)

withdrawn from the tip by the inappropriate passage of positive current) or to dye leakage, which appears as rapid fading during or after intracellular staining consequent on poor impalement. Dye-staining can sometimes be successfully achieved by use of the 'buzz' facility (if fitted) of the microelectrode preamplifier or by causing 'ringing' of the microelectrode by excessive capacity neutralization. Avoid using a dye-injection facility, except to clear the microelectrode tip before impalement, because cell damage is almost inevitable in these delicate neurones.

Authentic dye-coupling of neurones (as opposed to inadvertent sequential penetration of two cells whilst attempting single impalement) in both myenteric and submucous plexuses is exceptionally rare (only three instances in over 2500 impalements in this laboratory).

7. Intracellular electrophysiological recording

7.1 Hints on good impalement with fluorescent dye-filled microelectrodes

(a) Always use microelectrodes that settle, i.e. return to baseline quickly after passage of large but brief depolarizing and hyperpolarizing current (especially with activation of 'dye inject/clean' facility of the microelectrode preamplifier. Test the electrode prior to touching the tissue. The technique for impalement is described in *Protocol 6*.

(b) Use a polarizing and an orange or red filter when inspecting the tissue in transmitted light. Adjust the lighting and condenser lens to provide the least contrast, which mimics a shallow depth of field.

(c) In transmitted light, look for darker, round areas; these are likely to be nuclei of neurones. Do not attempt to impale neurones whose unstained soma outline can be clearly seen in transmitted light: these are cells that have been damaged, most probably during the dissection.

Protocol 6. Impalement of neurones

1. Focus slightly more deeply than the most superficial surface of the ganglion before lowering the microelectrode tip towards the tissue, so that the tip will lie to one side of what is thought to be the nucleus. The cytoplasm is very thin above and below the nucleus. Note that the point of focus of the electrode tip in transmitted light is not the same as that in incident light (epi-illumination).

2. Pass a small hyperpolarizing (negative) current (0.02–0.05 nA for 20–40 ms at 10 Hz) through the microelectrode and balance the bridge amplifier. Set the oscilloscope voltage gain to 10 mV/div and the sweep speed to 10 msec/div.

3. Advance the microelectrode until it touches the tissue. This is recognized at first as an increase in the electrode resistance (bridge imbalance), then as a downward displacement of the voltage baseline.
4. Back off slightly, so that there is little or no baseline displacement, before giving a single gentle but sharp tap. It will be necessary to experiment with the micromanipulator to learn where best to tap.
5. Watch for sudden downward displacement of the baseline, indicating impalement, with or without the firing of an action potential. Large membrane potentials are likely to be associated with glial cells. Some excellent neuronal impalements are associated with a relatively slow development of a resting membrane potential; often this is accompanied by the appearance of an electrotonic potential with a long time-constant.
6. Reduce the amplitude of the current pulses as soon as possible and reduce the rate of pulses to 1 Hz.
7. Be prepared to wait for up to 20–30 min for the appearance of slow after-potentials in certain neurones. Large depolarizing current pulses passed soon after impalement are liable to dislodge the electrode from the cell.

Although the temperature in the vicinity of the recorded neurone should be 35–37 °C, the temperature in the rest of the preparation may be rather lower, if an immersion lens (Leitz × 25/0.6 saltwater or Zeiss × 40/0.75 water) is used. Below 33 °C, the time course of action potentials is noticeably greater.

7.2 Properties of enteric neurones

In order to be able to classify the cell from which records have been made, it is essential that as many of the following characteristics or properties be assessed, once stable intracellular recording has been achieved:

- membrane potential (resting or spontaneous depolarizations)
- input resistance; time constant of electrotonic potential
- fast or slow excitatory synaptic potentials (cholinergic and non-cholinergic) and inhibitory synaptic potentials (adrenergic and non-adrenergic) in response to oral, anal, or circumferential focal or transmural stimulation
- rapid under-shoot and slow after-potential following generation of a somal action potential; the latter may be induced by antidromic invasion of the soma by stimuli delivered through the transmural or focal electrodes
- threshold and number of action potentials generated during a long (100–500 msec) depolarizing pulse

7.3 Pitfalls in recording and neuronal classification

From a review of the early literature, the impression would be gained that investigators had anticipated there being only a small number of electrophysiological type of neurone. It is clear, however, that enteric neurones are capable of exhibiting long-lasting changes in excitability that could affect the classification of an individual neurone.

Students might care to test the hypothesis that the same types of neurone are to be found throughout the gastrointestinal tract but that the proportions differ considerably between regions. Further, they could confirm that neurones of the submucosa undoubtedly have properties that are seldom seen in myenteric plexus neurones. Recent evidence strongly suggests that one can no longer take for granted the notion that the neuronal types in one location are typical for another, even within either the small or the large intestine (44–50).

7.3.1 S neurones and AH neurones

At present, there seems to be no perfect electrophysiological classification system. It is important that, whenever possible, mistakes are not made through inadequate examination of the characteristics of individual cells. If the system of Hirst *et al.* (3) is used, S neurones (*Figure 6*) should be restricted to enteric neurones with the following classical properties:

- clearly observable fast excitatory synaptic potentials (EPSPs) *and* no persistent after-hyperpolarization (slow AH) or increased membrane conductance lasting up to 30 sec, following the generation of a single soma action potential

the latter phenomena are the hall-marks of the so-called AH neurone (*Figure 6*), which by definition does not show fast EPSPs.

It is important to appreciate, however, that S neurones may show a persistent (0.5–3 sec) hyperpolarization following repetitive firing and that the classical AH (associated with a calcium-activated potassium conductance change) may be suppressed during a slow EPSP (induced by focal or transmural electrical stimulation), focal application of certain drugs or apparently spontaneously.

7.3.2 Other categories of neurone

Undoubtedly, however, there are neurones showing (a) both fast EPSPs and a slow AH (S/AH neurones), or (b) neither of these properties (no S/no AH neurones). The former type of neurone appears to be common in the myenteric plexus of large intestine and in the stomach (44–47). In such neurones, the slow AH may not be as persistent (only a few seconds), whereas the latter type of neurone may or may not be capable of maintained,

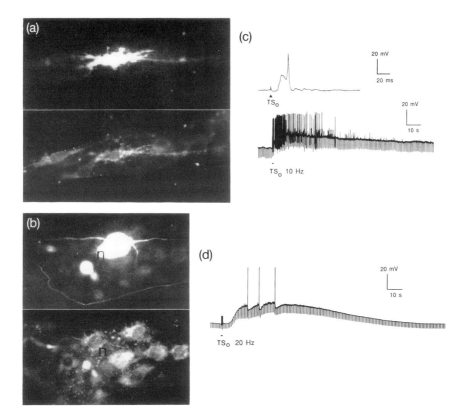

Figure 6. Cross-correlations between morphology, electrophysiology, and immunohistochemistry of identified guinea-pig myenteric plexus neurones. (a, b) Lucifer Yellow-stained neurones (upper pictures) and corresponding field of view (lower pictures) in preparations showing immunoreactivity for the opioid peptide, [Met^5]enkephalyl-Arg^6-Gly^7-Leu^8. In (a) note the similarity between Lucifer Yellow-staining and distribution of immunoreactivity in the Dogiel Type I cell, and in (b) note the lack of immunoreactivity in the soma and processes of the Dogiel Type II cell (lower picture); n, the position of the nucleus. (c, d) Intracellular electrophysiological recordings. (c) The traces are from the Dogiel Type I cells shown in (a), and the trace in (d) is from the Type II cell shown in (b). In the Type I cell, fast EPSPs, one of which was suprathreshold, were evoked by a single stimulus applied through the transmural electrode about 5 mm oral to the cell (TS$_o$). The full amplitude of the action potential was not faithfully reproduced by the pen recorder. Direct soma depolarization did not lead to the development of a long-lasting hyperpolarization. Thus, the cell is an S neurone. A slow EPSP was evoked in the same cell by repetitive stimulation (10 Hz for 1 sec) at the same intensity and location (lower trace). (d) Slow EPSP evoked by repetitive transmural stimulation (20 Hz for 1 sec; TS$_o$) about 3 mm from a cell which showed a marked long-lasting hyperpolarization following the soma action potential. Note the long-lasting change in input resistance during the slow EPSP, which brought the membrane potential to above threshold for the generation of action potentials. Fast excitatory synaptic potentials were not recorded in this neurone in response to either transmural stimulation or to focal stimulation applied to interganglionic fibre tracts and the surface of the ganglion near the soma. Thus, the cell is an AH neurone; such cells do not express opioid peptides (10, 18). (Figure courtesy of G. T. Pearson and G. M. Lees (unpublished observations).)

repetitive discharge of action potentials during a long depolarizing current pulse (100–500 msec), as shown by the classic S neurone (2, 3). Neurones that fire only one action potential in response to a depolarizing current pulse have been described and may be more common than previously acknowledged. Neurones showing fast EPSPs but incapable of generating action potentials have been reported to occur in both small and large intestine (see reference 10).

7.3.3 Slow EPSPs and slow inhibitory potentials

Although, in the myenteric plexus, only a small population of S neurones show muscarinic receptor-mediated slow EPSPs (51), non-cholinergic slow EPSPs are extremely common in both S and AH neurones (14, 21, 52, 53; *Figure 6*) but their nature and time course are probably not identical (54). It should be appreciated, however, that long lasting depolarizations cannot always be assumed to be slow EPSPs. It is necessary to verify this by showing that the depolarization is actually associated with an increased membrane excitability. Although slow excitatory events can be detected as a depolarization, the voltage change may be scarcely discernible at certain membrane potentials. The underlying *decrease* in membrane conductance is, however, more readily apparent. This is monitored as an increase in the amplitude of transient membrane potential changes induced by the passage of constant current pulses across the cell membrane, through the microelectrode. By Ohm's Law, the voltage changes reflect increases in the input resistance of the neurone; such changes, which may persist for 1–3 min in response to brief tetanic stimulation, often outlast the depolarization itself. Other slow depolarizations in myenteric plexus neurones are associated with such an *increase* in membrane conductance that the cell does not reach threshold for the firing of an action potential. Thus, this type of response is actually an inhibitory potential (55). Hyperpolarizing (adrenergic) inhibitory synaptic potentials (IPSPs) can be recorded in about 50% of submucous neurones (4, 56–58) and non-adrenergic (IPSPs) have also been described. In myenteric plexus neurones, IPSPs are rare, but biphasic events lasting many seconds may be recorded in a small proportion of neurones (59, 60).

7.3.4 Need for meticulous observation

Thus, it will be apparent that meticulous observations have to be made if meaningful correlations are to be made with morphology and neurochemical content (*Figure 6*). Finally, if a neurone appears to develop fast EPSPs or a slow AH, where none could be elicited after stable recording conditions had been established, check that only one neurone has been impaled. Under such circumstances, it is likely to become apparent that the tip of the microelectrode has moved and has passed into an adjacent cell, which has now also been stained. It should be noted that a change in resting membrane potential may not be apparent, since the second impalement is likely to have occurred during repetitive stimulation of the preparation, for example to elicit a slow EPSP.

Acknowledgements

This work was supported by the Medical Research Council, SmithKline (1982) Foundation, and the University of Aberdeen. The helpful comments of Dr A. D. Corbett and Dr J.-P. Timmermans are gratefully acknowledged.

References

1. Dogiel, A. S. (1899). *Arch. Anat. Physiol. Leipzig, Anat. Abteil* (Jg 1899) 130.
2. Nishi, S. and North, R. A. (1973). *J. Physiol.*, **231**, 471–491.
3. Hirst, G. D. S., Holeman, M. E., and Spence, I. (1974). *J. Physiol.*, **236**, 303.
4. Hirst, G. D. S. and McKirdy, H. C. (1975). *J. Physiol.*, **249**, 369.
5. Hodgkiss, J. P. and Lees, G. M. (1983). *Neuroscience*, **8**, 593.
6. Mihara, S., Katayama, Y., and Nishi, S. (1985). *Neuroscience*, **16**, 1057.
7. Furukawa, K., Taylor, G. S. and Bywater, R. A. R. (1986). *J. Neurophysiol.*, **55**, 1395.
8. Furness, J. B. and Costa, M. (1987). *The Enteric Nervous System*. Churchill-Livingstone, Edinburgh.
9. Costa, M., Furness, J. B., and Llwellyn-Smith, I. J. (1987). In: *Physiology of the Gastrointestinal Tract* (ed. L. R. Johnson), pp. 1–40. Raven Press, New York.
10. Wood, J. D. (1989). In: *Handbook of Physiology, Section 6. The Gastrointestinal System, Vol. 1. Motility and Circulation, Part 1*, pp. 465–517. American Physiological Society, Bethesda.
11. Bornstein, J. C., Costa, M., Furness, J. B., and Lees, G. M. (1984). *J. Physiol.*, **351**, 313–325.
12. Erde, S. M., Sherman, D., and Gershon, M. D. (1985). *J. Neurosci.*, **5**, 617–633.
13. Bornstein, J. C., Costa, M., and Furness, J. B. (1986). *J. Physiol.*, **381**, 465–482.
14. Katayama, Y., Lees, G. M., and Pearson, G. T. (1986). *J. Physiol.*, **378**, 1–11.
15. Bornstein, J. C., Costa, M., and Furness, J. B. (1988). *J. Physiol*, **398**, 371–390.
16. Bornstein, J. C. and Furness, J. B. (1988). *J. Auton. Nerv. Syst.*, **25**, 1–13.
17. Furness, J. B., Llewellyn-Smith, I. J., Bornstein, J. C., and Costa, M. (1988). In: *Handbook of Chemical Neuroanatomy, Vol. 6: The Peripheral Nevous System* (ed. A. Björklund, T. Hökfelt, and C. Owman), pp. 161–218. Elsevier, Amsterdam.
18. Iyer, V., Bornstein, J. C., Costa, M., Furness, J. B., Takahashi, Y., and Iwanaga, T. (1988). *J. Auton. Nerv. Syst.*, **22**, 141–150.
19. Stach, W. (1989). In: *Nerves and the Gastrointestinal Tract* (ed. M. V. Singer, and H. Goebell), pp. 29–45. MTP Press Limited, Lancaster.
20. Ekblad, E., Håkanson, R., and Sundler, F. (1989). In: *Nerves and the Gastrointestinal Tract* (ed. M. V. Singer and H. Goebell), pp. 47–56. MTP Press Limited, Lancaster.
21. Lees, G. M., Leishman, D. J., and Pearson, G. T. (1989). In: *Nerves and the Gastrointestinal Tract* (ed. M. V. Singer and H. Goebell), pp. 79–86. MTP Press Limited, Lancaster.

22. Furness, J. B., Morris, J. L., Gibbins, I. L., and Costa, M. (1989). *Annu. Rev. Pharmacol. Toxicol.*, **29**, 289.
23. Domoto, T., Bishop. A. E., Mitsuru, O., and Polak, J. M. (1990). *Gastroenterology*, **98**, 819.
24. Bornstein, J. C., Furness, J. B., Smith, T. K., and Trussell, D. C. (1991). *J. Neurosci.*, **11**, 505.
25. Brookes, S. J. H., Steele, P. A., and Costa, M. (1991). *Neuroscience*, **42**, 863.
26. Furness, J. B. and Bornstein, J. C. (1991). In: *Textbook of Gastroenterology* (ed. T. Yamada), pp. 2–24. Lippincott, Pennsylvania.
27. Schabadash, A. (1930). *Z. Zellforsch. mikroskop. Anat.*, **10**, 320.
28. Gunn, M. (1968). *J. Anat.*, **102**, 223.
29. Stach, W. (1977). *Verh. Anat. Ges.*, **71**, 867.
30. Stach, W. (1977). *Z. micrkroskop. Anat. Forsch.*, **91**, 737.
31. Christensen, J. and Rick, G. A. (1987). *J. Auton. Nerv. Syst.*, **21**, 223.
32. Hoyle, C. H. V. and Burnstock, G. (1989). *J. Anat.*, **166**, 7.
33. Timmermans, J.-P., Scheuermann, D. W., Stach, W., Adriaensen, D., and De Groodt-Lasseel, M. H. A. (1990). *Cell Tissue Res.*, **260**, 367.
34. Gabella, G. (1969). *Experientia*, **25**, 218.
35. Steele, P. A. and Costa, M. (1990). *Neuroscience*, **38**, 771.
36. Messenger, J. P. and Furness, J. B. (1990). *Arch. Histol. Cytol.*, **53**, 467.
37. Costa, M., Buffa, R., Furness, J. B., and Solcia, E. L. (1980). *Histochemistry*, **65**, 157.
38. Polak, J. M. and Van Noorden, S. (1987). *An Introduction to Immunocytochemistry: Current Techniques and Problems.* Microscopy Handbooks 11, Royal Microscopic Society. Oxford University Press.
39. Thomson, D. J. and Bradbury, S. (1987). *An Introduction to Photomicrography.* Microscopy Handbooks 13. Royal Microscopical Society. Oxford University Press.
40. Stewart, W. W. (1978). *Cell*, **14**, 741.
41. Lees, G. M. and Gray, M. J. (1982). *Scand. J. Gastroenterol.*, **17** (Suppl. 71), 169.
42. Furness, J. B., Bornstein, J. C., and Trussell, D. C. (1988). *Cell Tissue Res.*, **254**, 561.
43. Miller, J. P. and Selverston, A. I. (1979). *Science*, **206**, 702.
44. Wade, P. R. and Wood, J. D. (1988). *Am. J. Physiol.*, **254**, G522.
45. Wade, P. R. and Wood, J. D. (1988). *Am. J. Physiol.*, **255**, G184.
46. Schemann, M. and Wood, J. D. (1989). *J. Physiol.*, **417**, 501.
47. Schemann, M. and Wood, J. D. (1989). *J. Physiol.*, **417**, 519.
48. Pearson, G. T., Reekie, F. M., Maitland, V., Wilson, J. M., Leishman, D. J., and Lees, G. M. (1991). *J. Auton. Nerv. Syst.*, **33**, 189.
49. Leishman, D. J., Katayama, Y., and Lees, G. M. (1991). *J. Physiol.*, **438**, 347P.
50. Mackenzie, G. M. and Lees, G. M. (1991). *Regul. Peptides*, **53**, 244.
51. North, R. A. and Tokimasa, T. (1982). *J. Physiol.*, **333**, 151.
52. Hodgkiss, J. P. (1981). *Pflüger's Arch.*, **391**, 331.
53. Hodgkiss, J. P. and Lees, G. M. (1984). *Neuroscience*, **11**, 255.
54. Bornstein J. C., North, R. A., Costa, M., and Furness, J. B. (1984). *Neuroscience*, **11**, 723.
55. Hodgkiss, J. P. and Lees, G. M. (1986). In: *Autonomic and Enteric Ganglia:*

Transmission and its Pharmacology (ed. A. G. Karczmar, K. Koketsu, and S. Nishi), pp. 369–405. Plenum, New York.
56. Surprenant, A. (1984). *J. Physiol.*, **351**, 343.
57. Surprenant, A. (1984). *J. Physiol.*, **351**, 363.
58. North, R. A. and Surprenant, A. (1985). *J. Physiol.*, **358**, 17.
59. Johnson, S. M., Katayama, Y., and North, R. A. (1980). *J. Physiol.*, **301**, 505.
60. Johnson, S. M., Katayama, Y., Morita, K., and North, R. A. (1980). *J. Physiol.*, **320**, 175.
61. Pearson, G. T. (1986). *J. Physiol.*, **378**, 100P.
62. Pearson, G. T., Gray, M. J., and Lees, G. M. (1986). *Regul. Peptides*, **15**, 188.

PART III
Central nervous system

8

Whole-cell patch-clamp recording from neurones in spinal cord slices

TONY E. PICKERING, DAVE SPANSWICK, and STEVE D. LOGAN

1. Introduction

The development of the patch-clamp methodology in the laboratories of the 1991 Nobel prize winners Neher and Sakmann has revolutionized our understanding of ion channels. The patch-clamp recording technique was originally developed to allow high-resolution current recording from single ion channels (1). The vital step in this process is the formation of a gigaohm seal between the electrode glass and the membrane. This tight seal electrically isolates the few square microns of membrane in the pipette by reducing the leakage current to a level where picoampere currents flowing through single channels are resolvable.

Tight-seal recordings are obtained by positioning a blunt-ended (1–2 μm diameter), low-resistance (1–10 MΩ) micropipette against a membrane and applying gentle suction to the pipette interior. If conditions are favourable, then a gigaohm seal will rapidly form between pipette and membrane, isolating a patch. By rupturing this patch, whole-cell access is established, allowing recordings to be made of the current flowing across the entire cell membrane (whole-cell recording (WCR)) (1–3). This is analogous to conventional intracellular recording but offers advantages including high resolution, low noise, excellent stability, and control over intracellular constituents. These improvements are due to the low electrode resistance and its tight membrane–glass seal.

The advantages of WCR are particularly attractive to neuroscientists studying the small but electrically volatile vertebrate neurones. However, the requirement for visualization and cleanliness when using patch-clamp recording restricted its application to dissociated neuronal cultures. These cultures have the disadvantage that they are taken from immature tissue. Their subsequent maturation *in vitro* is not necessarily representative of the *in vivo* adult neurone (see Chapter 1).

This problem of uncertain differentiation has been solved by the method of acutely dissociating adult neurones (4). Enzymatically digested brain slices are gently triturated to isolate neuronal soma with proximal dendrites. The cells remain healthy for periods of up to 8 h. They have the advantage of being anatomically compact and, therefore, good for voltage-clamping. However, these isolated neurones suffer from both mechanical and enzymatic damage, and the loss of synaptic inputs.

An elegant organotypic culture technique allows explant cultures to be prepared from mammalian brain (5). This technique uses a roller culture technique to generate a monolayer from a tissue slice that retains much of the synaptic organization of the original explant. These cultures have proved ideal for patch-clamp recording as cells can be visually identified and synaptic responses studied.

The first WCR from slices were obtained from amacrine cells in the salamander retina (6). The retina is cut into 150 μm slices, which are relatively free from myelination, allowing cells to be identified with a × 40 water immersion lens. The first WCR from mammalian slices used Nomarski optics to visualize neurones in sections 100–140 μm thick (7). The tissue overlying the soma is first blown clear with a jet of saline from a wide-bore pipette. The exposed soma is then patched and WCR established. Thin slices have the disadvantage that neuronal dendritic trees can be severely truncated. However, conventional slices are too thick to allow the use of Nomarski optics.

A method for obtaining WCR from conventional slices without the need for visualization has been developed ('blind' WCR) (8). The patch-pipette resistance is monitored as it is lowered through the slice; an increase in resistance indicates contact with a cell (from which WCR may be obtained). High success rates are achieved using this technique, finally putting to rest the dogma of the need for absolute cleanliness when patching.

It is obvious from these and other studies that neuronal properties recorded with patch-clamp electrodes show marked differences from those seen with microelectrodes. The most notable differences are in input resistance and membrane time constant. WCR shows an order of magnitude increase in both parameters. Note, however, that most WCR studies are done at room temperature, which will also tend to increase these parameters. Intracellular penetration of cells produces a traumatic leakage pathway which is practically absent in WCR, hence the measured input resistance and time constant reflect those of the intact cell.

To date, the use of WCR in slices has allowed:

- recordings from cells too small to allow stable intracellular recording, e.g. cerebellar granule cells, somata less than 5 μm in diameter (9)
- recordings from identified neurones in heterogeneous brain regions by including non-toxic dyes in the pipette

- the observation of spontaneous miniature postsynaptic potentials in the central nervous system, e.g. GABAergic inhibitory postsynaptic potentials (IPSPs) in the dentate gyrus (7)
- examination of the mechanism of induction and the locus of enhancement in long-term potentiation (see reference (10) for review)
- study of synaptic transmission between pairs of neurones, using WCR to hold a postsynaptic cell while an intracellular electrode searches for presynaptic cells (11)
- calcium-imaging in neurones by including fura-2 in the patch pipette, e.g. pituitary secretory terminals (12)

We use WCR to examine sympathetic preganglionic neurones (SPN) in rat spinal cord slices. The majority of these cells are located in the lateral horn of the spinal cord. One of the main advantages of WCR in this study has been the ability to fill cells non-toxically with dyes. This has allowed unambiguous identification of recorded cells as SPN.

2. Spinal cord slice preparation

Spinal cord slices are cut from Sprague–Dawley rats (7–18 days old) of either sex. The whole procedure from removal of the cord to completion of slicing should take about 30 min (*Protocol 1*). This procedure requires Home Office approval by licence.

Protocol 1. Removing the spinal cord

Materials

- dissecting instruments: toothed forceps, scissors, iridectomy scissors, fine blunt forceps, 2 pairs watchmakers' forceps
- two Petri dishes, one containing a filter paper
- chilled (~4 °C) aerated artificial cerebrospinal fluid (aCSF) (*Table 1*)

Method

1. Place the rat on an anaesthetic machine and induce deep anaesthesia with a 6% Enflurane in oxygen mix. Check the depth of anaesthesia by monitoring the corneal blink and pedal withdrawal reflexes.
2. Secure the animal in a prone position with its shoulders raised over a bar beneath the chest. Using the scissors, cut away the skin overlying the vertebral column from the sacrum to the neck.
3. Reflect upwards the vascular fat pad (hibernating gland) found at the nape; take care to avoid excessive bleeding. Cut the superficial muscles from their attachments on the vertebral column.

Protocol 1. Continued

4. Incise the spinal column just above the sacrum, and decapitate the animal. This decapitation drains blood from the vessels surrounding the cord and makes the following dissection easier.
5. Grasp the dorsum of the lumbar spinal column with the toothed forceps and perform a laminectomy using the iridectomy scissors. This exposes the spinal cord from the low lumbar to the cervical region.
6. Bisect the cord at its lumbar end and gently lift it with the fine blunt forceps. Cut the ventral roots with the iridectomy scissors. Free the cord to the neck and place it in a Petri dish containing chilled aCSF.
7. Remove the dura by inserting the iridectomy scissors between it and the cord and cutting. Although it is possible to slice the cord at this stage, the pia acts like a sausage skin and tends to cause compression of the cord during slicing.
8. To remove the pia, hold the cord at its lumbar end with fine forceps and strip the pia upwards with a second pair. Usually a strip of pia is removed from each quadrant of the cord, along with the rootlets. For experiments requiring roots, it is possible to take single midline strips from the ventral and dorsal surfaces, leaving the rootlets intact.
9. Remove the cord from the aCSF and place it on a filter paper moistened with aCSF. Trim the cord with a razor blade to obtain an 8–10 mm thoracolumbar segment.

Table 1. Normal aCSF composition

Salt	Concentration (mM)
NaCl	127
KCl	1.9
KH_2PO_4	1.2
$CaCl_2$	2.4
$MgSO_4$	1.3
$NaHCO_3$	26
D-glucose	10

Our method of preparing slices is broadly similar to that documented by Takahashi (13), except that we cut thicker slices from older rats (*Protocol 2*).

Protocol 2. Preparation of slices

Materials

- 75 ml of 1.5–2% agar in aCSF pre-boiled and cooled to ~35–40 °C

- a mould to set the cord in agar
- a Vibratome (Technical Products International Inc., Series 1000)
- 500 ml of chilled aCSF
- cyanoacrylic glue
- wide-bore Pasteur pipette (slice carrier)
- storage chamber filled with aerated aCSF at room temperature

Method

1. Place the cord in a rectangular mould half-filled with 1.5–2% agar in aCSF (35–40 °C). Top up the mould with agar and immerse it in 4 °C aCSF to set the block. Embedding the cord gives it better support when slicing than glueing it direct to an agar block.
2. Remove the block from the mould, trim it, and glue it to a supporting block of agar. Secure this vertically to the stage of the Vibratome, rostral end uppermost with either the dorsal or ventral surface facing the blade.
3. Immerse the tissue in cold, aerated aCSF which is replaced frequently to maintain the temperature close to 4 °C. Cut transverse slices of thickness between 350 and 600 μm using a slow advance speed and moderate vibration.
4. Typically, between 7 and 12 slices are obtained. We make no attempt to trim any agar from the slice perimeter. Transfer the slices to the storage chamber using the slice carrier and incubate for at least 1 h before transfer to the slice bath.

3. Maintenance of slices

The slice storage chamber consists of a plastic tea strainer balanced upon a 250 ml beaker filled with aCSF. The slices sit upon the mesh of the strainer and are fully submerged in aerated aCSF at room temperature. A modified Petri dish is used as a lid to reduce oxygen loss. We have maintained slices in this way for over 24 h before recording.

For recording, a single slice is transferred to the submersion chamber (see *Figure 1*). We also use this bath design for our intracellular studies. The slice sits upon a nylon grid (wedding veil) in the main chamber (*Figures 2* and *3*). It is secured with a slice holder, consisting of a second nylon grid stretched across a silver wire frame held in a manipulator. Aerated aCSF is conducted through the bubble trap into the main chamber where it is directed up over the slice and out into the drainage reservoir. A peristaltic pump is used to drain the aCSF.

The aCSF is perfused through the bath at 2–6 ml/min from a gravity-feed reservoir bottle. Most recordings have been made at room temperature,

Whole-cell patch-clamp recording from neurones

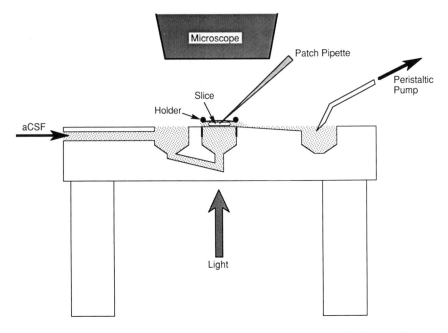

Figure 1. Schematic representation of the slice bath. The bath is drilled from a Perspex disk and raised upon cylindrical legs. The slice is held on a nylon mesh grid, with a second grid strung on a silver wire frame. Aerated aCSF flows into a bubble trap before being directed over the slice. It is removed using a peristaltic pump at a rate of 2–6 ml/min.

although we have also worked at 34–36 °C without difficulties. The aCSF in the reservoir bottle is continuously aerated with 95% O_2/5% CO_2 throughout the experiment. Drugs are added to the perfusate from a series of 20 ml syringe bodies placed upon three-way taps arranged in series with the main perfusion line.

Fresh aCSF is prepared on the day of the experiment. To avoid precipitation of calcium salts it is saturated with carbogen before the $CaCl_2$ is added. This saline has an osmolarity of ~318 mOsm and pH of 7.4.

4. Whole-cell recording from neurones in slices

4.1 The rig

The recording rig is similar to that used by us for intracellular recording studies. The slice is trans-illuminated with a fibre optic (Schott 1500-KLT), and the lateral horn identified under a binocular microscope. Trans-illumination enhances the contrast between grey and white matter in the slice. The pipette is mounted in the standard pipette holder provided with the

Figure 2. Photomicrograph of a slice chamber. This shows a spinal cord slice held in position between nylon grid as it is superfused with aCSF. The patch pipette is seen above the bath on the right. A puffer pipette, for localized application of drugs, and a bipolar stimulating electrode are also seen. In the background is the siphon leading to the peristaltic pump for removal of aCSF.

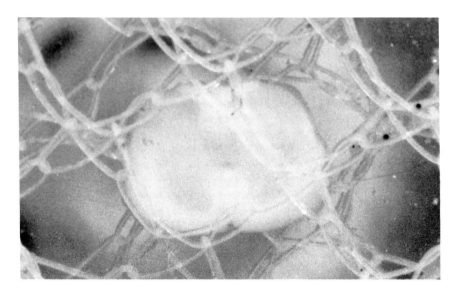

Figure 3. Photomicrograph of a spinal cord slice. A slice from a 14-day-old rat is seen *in situ* in the recording bath. Trans-illumination of the slice highlights the delineation between grey and white matter, allowing the lateral horn to be recognized.

amplifier (List-electronic EPC-7). Pressure is applied, by mouth, via a 1 ml syringe body connected by small-bore tubing to the holder. The electrolyte is contacted by a silver/silver chloride wire in the pipette holder. The holder is fixed in a custom built swivelling arm. A small length of wire links the holder to the headstage of the amplifier. This arrangement allows pipettes to be changed easily. It also keeps the bulky headstage away from the slice bath, making room for other manipulators. The swivelling arm is held by a 3-axis hydraulic microdrive (Narishige, MO-203) to allow fine manipulation of the pipette. The silver wire of the slice holder is coated with silver chloride and used as the bath earth.

Noise and hum (50 Hz) are reduced to a minimum by carefully earthing all conducting objects to the high-quality earth provided on the amplifier. The equipment is surrounded by a Faraday cage to further reduce interference. All output signals are filtered at 3 kHz using the internal filter in the amplifier. Output is observed on a digital oscilloscope (Gould 1602) and a two-channel chart recorder (Gould 2400S). All recordings are stored on videotape for offline analysis (using a Sony PCM 701, modified after reference 14).

To evoke postsynaptic potentials and antidromic spikes, a concentric bipolar stimulating electrode (SNE-100, Clark Electromedical) is positioned upon the surface of the slice. Typical stimuli range between 4 and 10 V and are applied for 0.1–1 msec from an isolated stimulator (Digitimer DS2) driven by a pulse generator (Digitimer 3290).

We use a pressure ejection unit (custom built) to allow focal drug applications from puffer pipettes (5–10 μm tip). Agonist solutions have concentrations between 50 μM and 1 mM. Fast green (0.005%) can be added to the drug solution to visualize the puff. Both the ejection pipette and the stimulating electrode are held in manipulators allowing them to be moved about the surface of the slice.

We digitize data into a PC-AT class computer using Sigavg software (an oscilloscope emulation) to drive a 1401 intelligent interface (CED). We use this software for straightforward analysis, such as current–voltage measurements, averaging, and figure production. For more complex data analysis such as fitting exponentials to charging curves, we transfer the data to an Apple Mac IIfx and import it into Systat (Systat Inc.). Here a generalised non-linear curve-fitting algorithm is used to fit functions to the data.

4.2 Patch pipettes

In early experiments pipettes were pulled in two stages on a vertical puller (Narishige PB-7) and lightly fire-polished using a microforge (Narishige MF-9). We now pull our pipettes on a Sutter P87 programmable horizontal puller, as no polishing is needed. These electrodes produce more consistent results and are easier to fabricate. They are pulled in two stages from borosilicate filamented thin-wall tubing (Clark Electromedical, GC150–

TF10). The electrodes are back-filled with pipette solution immediately before being used. When filled, they have resistances of 5–10 MΩ.

Over the past 2 years we have used a variety of glass types, capillary dimensions, and pull parameters to prepare patch pipettes of different shapes and resistances. We are still unable to tell how well a new pipette will perform before it is used. Many different patch electrode shapes will work, many other simply will not. Only trial and error will uncover good electrodes. However, there are several situations to avoid:

- make sure no bubbles remain in the pipette solution after filling
- ensure that saliva does not track down the suction tubing to form an air-lock
- with the electrode in the bath, application of positive pressure should not result in an increase in pipette resistance

4.3 Pipette solutions

We have designed a solution with the objective of approximating the original intracellular milieu (see *Table 2*). Several ionic concentrations have been derived from the reversal potentials see with intracllular recordings. An after-hyperpolarization (AHP) seen in SPN following a train of action potentials is mediated by a Ca^{2+}-activated potassium conductance. We have reversed this AHP at around −97 mV, suggesting an intracellular potassium concentration of approximately 140 mM. We have also examined a glycinergic response in SPN which reverses at −64 mV, suggesting an intracellular Cl^- of about 11 mM.

Initial experiments on hippocampal neurones with 110 mM KCl in the pipette solution showed chloride-dependent giant EPSPs which were blocked by bicuculline, suggesting that they were reversed $GABA_A$ IPSPs. In order to obtain a more realistic reversal potential for chloride we substitute potassium gluconate for KCl. We have also substitued methysulphate for gluconate without any apparent difference in the data obtained.

The condition of cells recorded with high-chloride solutions showed a decline once WCR was established. However, cells recorded with gluconate solutions usually remained stable for several hours. The run-down may be due to an osmotic imbalance caused by the combination of non-diffusible anions in the cell and diffusible permeant Cl^- in the pipette solution.

Table 2. The Nernst reversal potentials for main permeant ions

	Pipette (mM)	aCSF (mM)	Reversal potential (mV)
K^+	144	3.1	−98
Na^+	38	153	+35
Ca^{2+}	0.01	2.4	+156
Cl^-	16	133.7	−54

Gluconate has a larger molecular weight than Cl⁻ and, therefore, diffuses more slowly into the cell, perhaps making it less prone to creating an osmotic imbalance.

The early solutions were hypo-osmotic (~270 mOsm) in the belief that this aided patch formation and enhanced cell viability. Experience has shown that it is as easy to obtain WCR with solutions that are iso-osmotic with the aCSF (~318 mOsm). This has allowed us to increase the concentration of potassium gluconate to 130 mM to produce a more realistic K^+ reversal potential. Included in this pipette solution is 34 mM NaOH, used to titrate against the EGTA and Hepes. The presence of this Na^+ ensures that its reversal potential is within the physiological range.

Free calcium is buffered at 10^{-8} M with EGTA. The Ca^{2+}-activated potassium conductance-mediated AHP seen with intracellular recording is observed to run down within several minutes when using WCR. This probably indicates that the Ca^{2+}-buffering of the solution is too strong. However other Ca^{2+}-sensitive conductances, such as those involved in the AHP following a single spike, remain stable.

We include a physiological level of ATP (2 mM) in the pipette solution, along with 2 mM Mg^{2+}, which is probably almost completely bound to ATP. We have included GTP (0.5 mM) in the pipette solution when examining G-protein-coupled receptor responses. However, no rundown is evident in these receptor responses without GTP, so it is not routinely included.

Lucifer Yellow is added to the pipette solution to allow later identification of the cells (15). The low concentration of Lucifer Yellow needed to fill cells when using WCR allows the use of the potassium salt rather than the more soluble (and toxic) lithium salt. It is possible to kill cells which are full of Lucifer Yellow with strong illumination. This is due presumably to the generation of a phototoxic product of Lucifer Yellow. Care should therefore be taken to limit the exposure of cells to strong illumination.

The pipette solution (*Table 3*) is prepared from stocks on the day of the experiment (1 ml is usually sufficient). All salts are stored as 10 × final concentration stock solution except Lucifer Yellow which is stored as a 5 ×

Table 3. Standard pipette solution

Salt	Concentration (mM)
K gluconate	130
KCl	10
CaCl$_2$	1
Na$_3$-EGTA	11
Hepes	10
MgCl$_2$	2
Na$_2$ATP	2
K$_2$ Lucifer Yellow	2
pH = 7.4 with NaOH	

final concentration stock solution. The gluconate stock solution constitutes an attractive culture medium for microorganisms, therefore new stock solutions should be made regularly. The ATP, GTP, and Lucifer Yellow stock solutions are stored frozen.

We have also used biocytin (5–20 mM) in the patch solution in an attempt to uncover gap junction coupling between SPN (18). WCR was identical with Lucifer Yellow or biocytin, although neither has shown coupling. The fluorescence signal produced by conjugation with fluorescein-labelled avidin (25–50 μg/ml, Vector Laboratories) is less intense than that of Lucifer Yellow.

4.4 Obtaining whole-cell recordings

We use the 'blind' method of obtaining WCR (8) without prior visualization or cleaning of neurones (*Figures 4* and *5*). This description refers specifically to settings available on the EPC-7; however, other patch-clamp amplifiers will have equivalent controls (*Protocol 3*).

Protocol 3. Obtaining whole-cell recordings

1. Insert the filled pipette into the holder and position it above the bath aligned with the lateral horn of the slice. Check for the presence of small air bubbles close to the electrode tip. Remove any bubbles by applying strong suction before the pipette is in the bath saline. Apply positive pressure to the back of the pipette as it is lowered through the air–liquid interface.

2. Switch on the amplifier in search mode (gain 2 mV/pA) and correct any offset potentials (using Vp-offset). Apply small (1 mV × 25 msec, 5 Hz) depolarizing pulses to the pipette. Follow changes in pipette resistance as it is advanced into the slice by observing the current response to pulses on an oscilloscope at a fast sweep speed (5 msec/div).

3. Occlusion of the pipette tip is indicated by a decrease in the current pulse amplitude. Apply gentle positive pressure to clear the obstruction and restore the current pulse. The presence of a cell is indicated by a rebound decrease in the pulse amplitude when the pressure is released.

4. Advance the electrode slightly, which will further reduce the current pulse. Repeat the pressure application. The distance from the cell can be gauged by the speed of recoil following gentle pressure application (the closer the cell, the faster the recoil). The electrode is usually advanced until the current pulse has decreased to between 40 and 25% of its original amplitude.

5. Apply very gentle suction to form a cell-attached patch. The gigaohm-seal usually forms rapidly (< 15 sec) and may improve slightly over several minutes. If patch formation does not occur, do not suck harder as

Protocol 3. *Continued*

  you will only tear a patch from the cell! Withdraw the electrode slightly and then re-advance for a second attempt. If this fails, try a different cell or replace the electrode.

6. When a seal is obtained, switch the amplifier to voltage-clamp mode and increase the gain to 10 mV/pA. Cancel the fast capacitative transient using the c-fast control. By increasing the voltage step to 10 mV it is possible to measure the patch resistance (normally 2–10 GΩ). Cell-attached patches typically show outward single channel currents and sometimes extracellular spikes.

7. Take the holding potential down to around −30 mV before applying gentle suction to rupture the seal. Rupture is often a gradual process that begins with an increase in channel activity, accompanied by a slow positive drift in the current trace. This is followed by a sudden jump in the holding current, indicating rupture of the patch, and the appearance of large spike currents.

8. Hyperpolarize the cell further to approximately −60 mV. Estimate the access resistance by applying −10 mV steps and measure the capacitative current transients. These may be offset, for voltage-clamp studies, by using the c-slow and g-series controls. The access resistance can often be improved by pulling the electrode back slightly from the cell.

9. Switch to current-clamp mode. We take a resting potential of greater than −40 mV and action potentials that overshoot zero to be indicative of a healthy neurone.

With this technique, between 30 and 50% of all electrodes used produce whole-cell recordings. Outside-out patches are frequently formed when the electrode is withdrawn from the cell. For unique identification of cells, only one neurone per slice is recorded from.

When trying to find neurones several points are worth noting:

(a) It is not necessary to apply constant positive pressure to the pipette to keep it clear while it is in the slice.

(b) If cells are not encountered within 150 µm into a slice, it is worth checking that the electrode is tracking through the grey matter.

(c) Care is needed when advancing electrodes onto cells; if too much force is used they tend to expire in a burst of spikes!

(d) If is possible to obtain recordings from cells deep in slices by blowing clear and pushing past any obstructions until the required depth is reached.

(e) As SPN occur in rostro-caudal 'nests' separated by several hundred micrometres, when one cell is found there are often many more in the near vicinity.

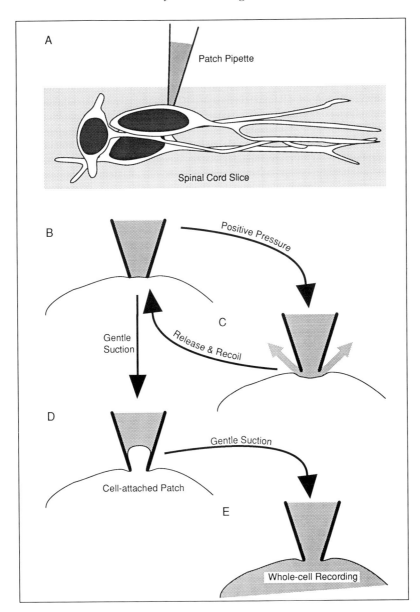

Figure 4. Schematic of procedure for obtaining 'blind' WCRs. Cells are detected by monitoring the pipette resistance as it is lowered through the slice. Cell contact occludes the pipette tip, causing an increase in pipette resistance (A and B). The application of gentle pressure removes the obstruction, restoring the pipette resistance (C). If the pipette is against a cell, then the resistance will increase when the pressure is released because of elastic recoil. We use the speed of recoil to estimate the closeness of pipette and cell. The application of gentle suction at this point will result in the formation of a cell-attached patch (D). This can be ruptured with further suction to produce a WCR (E).

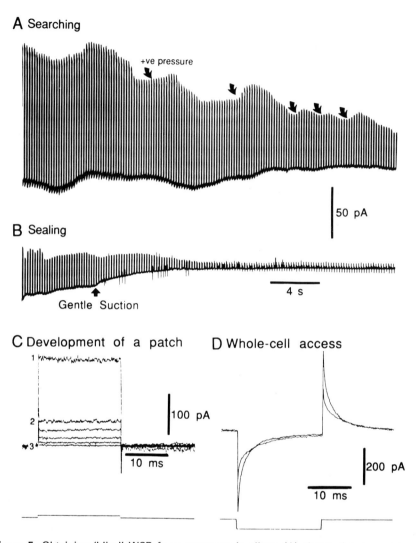

Figure 5. Obtaining 'blind' WCR from neurones in slices. (A) shows the current trace obtained as the patch pipette is moved through the slice; its resistance is monitored by applying small voltage pulses (1 mV × 20 msec). The pipette resistance increases when its tip encounters any obstruction. These obstructions can be cleared by applying gentle positive pressure (arrows). The presence of a cell is indicated by the elastic recoil after the pressure is released. The pipette is gradually advanced onto the cell. (B) Once the pipette is in the right position, the application of very gentle suction results in the formation of a cell-attached patch. (C) Shows this process on a faster time base. The initial pipette resistance is 4.5 MΩ (1), suction is applied at (2) and the patch is formed of resistance about 7 GΩ (3). (D) By applying further suction the patch is ruptured to yield WCR. This is indicated by the appearance of large capacitative transients (in response to a 10 mV step). The access resistance is improved in this cell by withdrawing the pipette slightly once the seal has ruptured (increasing the capacitative transient). This cell has an input resistance of approximately 550 MΩ.

(f) Neuronal somata are considerably larger than glial somata, therefore WCR is more commonly obtained from neurones. However, some glial cells are inevitably accessed. They are recognizable as they have a large resting potential (~ -70 mV), low input resistance and no sodium spikes.

A note of caution however. We mistook a population of relatively inexcitable SPN (with very high firing thresholds) as being glial cells for several months, before looking at their anatomy post-recording and realizing that they were neurones!

4.5 Perforated-patch recordings

A second method of obtaining whole-cell access is to include a pore-forming antibiotic, such as nystatin, in the pipette solution (16; see also Chapter 2). Once a cell-attached patch is obtained, the antibiotic incorporates into the membrane to perforate it. This perforated patch typically takes tens of minutes to form but can give low access resistances comparable to WCR. The major advantage of this technique is that intracellular dialysis is restricted because the pores are selective for monovalent ions and small molecules (i.e. smaller than glucose).

We have used this approach to examine the properties of SPN that have not been subjected to the insult of intracellular washout. We use a solution of the following composition (mM): potassium gluconate, 120; NaCl, 20; $MgCl_2$, 2; Hepes, 10; Lucifer Yellow, 5. As dialysis is limited, there is no need for Ca^{2+}-buffering or ATP in the pipette solution. Nystatin is added (from a frozen stock solution, 50 mg/ml in dimethylsulphoxide (DMSO)) at a concentration of 50–100 µg/ml and then sonicated. The solution is made up fresh before each experiment. We have experienced no additional difficulty in obtaining patches with nystatin in the pipette.

When this technique is used a Donnan equilibrium is set up across the perforated patch. Using the above solution, the intracellular equilibrium concentrations should be as follows: K^+ 108, Na^+ 36, Cl^- 44 (assuming 100 mM non-diffusible intracellular anions). Following recording it is possible to rupture the perforated patch and fill the cell with Lucifer Yellow for identification. However, this is less reliable than conventional WCR. Considerable variation in the degree of patch perforation was seen when using this technique. This may be due to the pipette tip picking up debris as it is moved through the slice, before patch formation. These debris may impair the access of nystatin to the patch of membrane.

5. Visualization of neurones

To allow cells to be identified after recording, Lucifer Yellow is included in the pipette solution (see *Figure 6*). This diffuses into the cell, producing a good fill in less than 5 min after obtaining WCR. After recording, slices are removed from the experimental bath and fixed in 10% formaldehyde in 0.1 M phosphate-buffered saline (PBS) for more than 6 h.

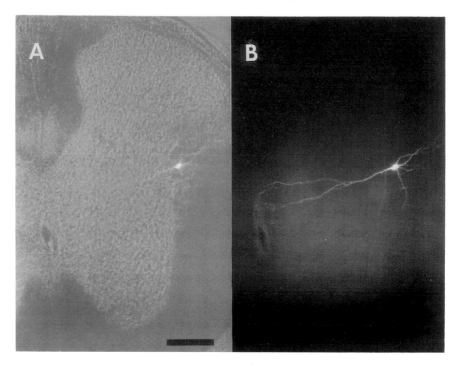

Figure 6. Morphology and identification of SPN. This figure shows a SPN filled with Lucifer Yellow in a slice from a 10-day-old rat viewed at low power. (Bar represents 200 µm). (A) the slice is both epi-illuminated and trans-illuminated to allow the position of the neurone to be determined. The cell soma is located in the lateral horn. (B) At the same magnification, without trans-illumination, the dendritic tree of this cell is seen. Note the pronounced mediolateral dendrites which project to an area just dorsal to the central canal. The axon of this cell can be seen to emerge from the medial dendrite and head towards the ventral horn. The soma has dimensions 27 × 15 µm.

Slices containing cells filled with biocytin are permeabilized after fixation by incubation for 1 h in a 1% solution of Triton X-100 in 0.1 M PBS. This is followed by 2–4 h in the same solution with fluorescein avidin 25–50 µg/ml, before being rinsed in PBS.

The slice is cleared by immersion in DMSO for about 30 min (17). This technique avoids the tissue shrinkage encountered with graded alcohol dehydration. The slices are mounted in a pool of DMSO between two coverslips sealed with paraffin wax. This method of mounting allows the slice to be flipped over to allow the neurone to be viewed from either side of the slice (through the least tissue).

Neurones can be viewed on an epifluorescence microscope with filters for Lucifer Yellow or fluorescein. We measure the cells using an eyepiece graticule and photograph them onto 35 mm film. One disadvantage of using fluorescent dyes is that they make *camera lucida* drawing difficult. By reacting an avidin-horseradish peroxidase conjugate against biocytin-filled neurones, it is possible to form an opaque reaction product ideal for producing such drawings (18).

We also use a laser scanning confocal microscope (Biorad, MRC-600) to examine cells. This allows neurones to be sectioned optically and then reconstructed using the image manipulation software available with the microscope. We typically take 15–20 sections of each cell at 1 μm intervals. These images are stored on disk and can be reconstructed and analysed at a later date. This method of image production gives excellent resolution. It also allows the generation of *camera lucida*-like 2-D projections of neurones. A further advantage is that it is possible to see clearly the rostro-caudal organization of neurones in transverse slices, which is normally difficult to assess.

6. Results

Using the techniques described above, we have obtained high-resolution recordings from over 200 identified spinal cord neurones (see reference 19). Analysis of cell morphology post-recording has allowed us to discriminate between SPN and cells located in adjacent lamina (particularly motoneurones and cells in lamina III and VII). The SPN have exhibited a typical morphology with somata in the lateral horn, an axon which leaves the ventral horn, a large medial dendritic projection and often lateral and rostro-caudal dendrites (see *Figure 6*).

The recordings have lasted for up to 5 h and show exceptional stability over this period, facilitating pharmacological studies. The SPN exhibit clearly different electrophysiological properties from other spinal cord neurones. The SPN recorded using WCR exhibit similar properties to those we have seen by use of conventional intracellular methods. However, we have seen that when WCR is used, the input resistances and time constants of these cells are considerably larger (see *Figure 7*). We believe this increased input resistance (and time constant) is due to the absence of the microelectrode penetration leakage conductance. We see similar time constants when using perforated patch recording, which suggests that the increase is not due to intracellular dialysis. A comparison between data obtained in our laboratory is shown below (*Table 4*). It is interesting to note the large range of resting potentials and input resistances seen in SPN recorded using the same pipette solution; this emphasizes the importance of resting membrane conductances in determining the electrophysiological properties of the neurones.

7. Prospects

Using WCR from slices, we have been able to make high-quality electrophysiological recordings from identified spinal cord neurones. Neuronal

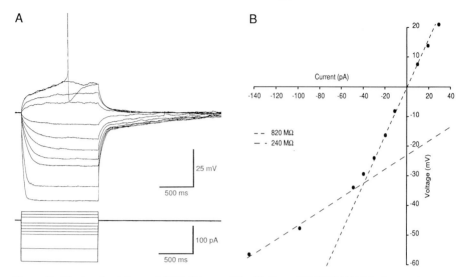

Figure 7. Electrophysiology of SPN. ($V_m = -58$ mV, $R_{in} = 820$ MΩ) (A) This figure shows the voltage responses of the cell seen in *Figure 6* to injected current pulses. Injection of depolarizing current pulses brings the neurone to threshold, resulting in spike firing (threshold = 20 mV, amplitude = 59 mV, duration = 8.1 msec). Hyperpolarizing pulses show clear inward rectification, and also activate a transient rectification on their offset. (B) Data plotted as a current–voltage relationship. The input resistance at rest is 820 MΩ; also illustrated is the dramatic rectification seen at potentials greater than 25 mV below rest.

identification is particularly important when working with a non-homogeneous slice of tissue such as the spinal cord, and also when working with a functionally diverse group of neurones such as SPN. This identification can be taken a stage further by retrogradely prelabelling neurones *in vivo* before preparing slices (see references 20 and 21). With either a conventional fluorescence microscope or a laser scanning confocal microscope, it will be possible to record from prelabelled neurones with the objective of linking a function with the recorded properties. A further extension to this will be the advent of *in*

Table 4. Comparison between intracellular and whole-cell recordings from SPN. Figures in brackets show range

	Intracellular		Whole-cell	
	(35 °C, $n = 20$)		(20 °C, $n = 32$)	
Resting potential (mV)	−61	(−51 to −75)	−52	(−40 to −82)
Input resistance (MΩ)	128	(46–190)	950	(340–2800)
Time constant (ms)	23	(6–56)	118	(31–360)
Threshold (mV)	25	(15–46)	17	(4–45)
Spike amplitude (mV)	55	(43–67)	66	(47–83)
Spike duration (msec)	3.1	(1.8–4.8)	8.8	(4.0–14.5)
After-hyperpolarization (mV)	21	(15–27)	24	(14–33)

vivo WCR. The advantages of excellent mechanical stability and dye-filling of cells provided by WCR may partly reverse the trend for *in vitro* recordings in favour of *in vivo* preparations.

It is already clear that the increased resolution provided by WCR will give us many new insights into neuronal function. However, intracellular recording is by no means obsolete. The problems associated with the intracellular washout of cells by the pipette contents will ensure intracellular recording has a role for some time to come.

Acknowledgements

Much of the work described herein was carried out by A. E. P. whilst in receipt of a Wellcome Trust prize studentship. We would like to thank Ian Gibson and Hamish Ross for their contributions to this work. We are additionally indebted to Trevor Hayward for his photographic expertise and to Dr Gerard Johnson for introducing us to confocal microscopy. We also applaud our technical workshop staff for their help in constructing the rigs. This work is supported by grants to S. D. L. from the British Heart Foundation and the MRC.

References

1. Hamill, O. P., Marty, A., Neher, E., Sakmann, B., and Sigworth, F. J. (1981). *Pflügers Arch.*, **391**, 85.
2. Fenwick, E. M., Marty, A. and Neher, E. (1982). *J. Physiol.*, **331**, 577.
3. Sakmann, B. and Neher, E. (ed.) (1983). *Single Channel Recording*. Plenum Press, New York.
4. Kay, A. R. and Wong, R. K. S. (1986). *J. Neurosci. Methods*, **16**, 227.
5. Gahwiler, B. H. and Knopfel, T. (1990). In *Preparations of the Vertebrate Central Nervous System In Vitro*, Vol. **13** (ed. H. Jahnsen), pp. 77–102. John Wiley, Chichester.
6. Barnes, S. and Werblin, F. (1986). *Proc. Natl. Acad. Sci.*, **83**, 1509.
7. Edwards, F. A., Konnerth, A., Sakmann, B., and Takahashi, T. (1989). *Pflügers Arch.*, **414**, 600.
8. Blanton, M. G., LoTurco, J. J., and Kriegstein, A. R. (1989). *J. Neurosci. Methods*, **30**, 203.
9. D'Angelo, E., Rossi, P., and Garthwaite, J. (1990). *Nature*, **346**, 467–470.
10. Edwards, F. A. (1991). *Nature*, **350**, 271.
11. Malinow, R. (1991). *Science*, **252**, 722.
12. Jackson, M. B., Konnerth, A. and Augustine, G. J. (1991). *Proc. Natl. Acad. Sci.*, **88**, 380.
13. Takahashi, T. (1990). *J. Physiol.*, **423**, 27.
14. Lamb, T. D. (1985). *J. Neurosci. Methods*, **15**, 1.
15. Stewart, W. W. (1978). *Cell*, **14**, 741.
16. Horn, R. and Marty, A. (1988). *J. Gen. Physiol.*, **92**, 145.

17. Grace, A. A. and Llinas, R. (1985). *Neuroscience*, **16**, 461.
18. Horikawa, K. and Armstrong, W. E. (1988). *J. Neurosci. Methods*, **25**, 1.
19. Pickering, A. E., Spanswick, D., and Logan, S. D. (1991). *Neurosci. Lett.*, **130**, 237.
20. Katz, L. C. (1987). *J. Neurosci.*, **7**, 1223.
21. Viana, F., Gibbs, L. and Berger, A. J. (1990). *Neuroscience*, **38**, 829.

9

The spinal cord as an *in vitro* preparation

JEFFERY BAGUST

1. Introduction

The last decade has seen an explosive increase in the use of isolated mammalian nervous system preparations for the study of the biophysics, physiology, and pharmacology of the central nervous system (CNS) (1, 2, 3). They offer to the mammalian brain researcher advantages that previously were only available to those working on isolated invertebrate nervous systems:

- absence of anaesthetic agents
- control over the composition of the medium surrounding the tissue, allowing known concentrations of test compounds to be added
- improved mechanical stability; movements caused by respiration and blood pressure are eliminated
- control over activity entering the system; inputs from other brain areas and the peripheral nervous system are removed
- improved visibility of structures within the preparation allows electrodes to be guided accurately by visual means to the recording site
- experimental convenience; once isolated the tissue is classified as a preparation and is not subject to the legislation that controls the use of living animals in experiments

Most *in vitro* CNS preparations involve thin slices of tissue (200–400 μm) to enable adequate oxygenation of the cells in the middle of the slice by diffusion from the surface. These preparations, however, have several disadvantages over the *in vivo* condition, of which the foremost is the interruption of long pathways, both within the slice, and into and out from the slice from other brain structures and the peripheral nervous system.

The isolated mammalian spinal cord preparation has been an exception in that the more usual preparation has been a hemisected or whole cord, with thin transverse slice preparations being a less common, but very useful, technique where activity at the segmental level is to be investigated. The hemisected or whole cord retains clearly identified input and output pathways in the dorsal and ventral roots, which, if long enough, can be isolated from the cord itself and be used to provide controlled sensory input to the system. In addition, the longitudinal tracts within the cord are retained, allowing the spread of activity over many spinal segments to be investigated, and in the whole cord, activity crossing the midline can also be studied. It has proved possible to dissect the cord out in continuity with peripheral structures such as muscles and skin, and isolated spinal cord/tail or spinal cord/skin/muscle preparations promise to make valuable contributions in the near future to our understanding of spinal physiology (see also Chapter 12).

Most of the isolated spinal cord preparations to date have used material taken from neonatal animals. This has the advantage that the spinal cord is easily dissected, the bone being in an early state of calcification, and there is little connective tissue. It is also highly resistant to hypoxia, staying viable for many hours or even days. However, such preparations are very immature, myelination is incomplete, the motoneurones fire prolonged calcium action potentials, and there is evidence that the pharmacology may not be typical of the adult condition. Behaviourally, neonatal animals are only capable of the simplest movements and it is probable that much of the circuitry in the spinal cord has yet to develop fully. In addition the spinal roots are short and can be difficult to isolate for stimulation and recording.

In this chapter I will describe a spinal cord preparation taken from more mature animals than the neonatal rat pups that are often used. These animals are fully mobile, able to fend for themselves, and their cords have adult properties. There is, of course, a penalty for this increased maturity; the cords are thicker and less resistant to anoxia than those from younger animals. Deep in the centre of the cord, the tissue is undoubtedly hypoxic and much of it is probably moribund. However, the dorsal horn and the ascending and descending tracts stay alive well, and under appropriate conditions good segmental reflexes can be obtained from the lumbar and sacral roots on stimulation of the dorsal roots.

As always, it is for the investigator to decide on the most appropriate preparation or combination of preparations for the investigation being undertaken. By describing in detail this preparation, which is one of the more difficult *in vitro* mammalian nervous system preparations, it is hoped to encourage further development in this field. Even the most experienced investigators may pick up a few tips to assist their work.

2. Species and age

Most workers on *in vitro* mammalian CNS preparations use material taken from the laboratory rat, and there have been reports of successful recordings from sacral segments of the cord taken from adult rats. But the thickness of the cord probably means that the use of adult rat tissue will be confined to small segments or thin slices of the cord in order to maintain viability. Mice provide a smaller alternative, reaching a weight of only 40–50 g in the adult, but experience has shown that it is difficult to evoke much activity in cords taken from mice weighing more than 15 g. The reason for this is not clear. Mice also have relatively short lumbar roots which can make them difficult to isolate for stimulation and recording.

The mammal from which the best activity has been obtained in the whole-cord preparation is the Syrian, or golden, hamster (*Mesocricetus auratus*). These animals are intermediate in size between the mouse and the rat, and their cords show excellent viability when isolated. It is not clear why hamster cords should be more viable than the smaller mouse cord: it may be related to their ability to hibernate, enabling their nervous tissue to function at a lower temperature, and in the presence of lower levels of oxygen. An additional advantage of using hamster cords is the length of the lumbar spinal roots. The golden hamster has a relatively short cord, similar to that of humans, which in the adult terminates in the region of the upper lumbar vertebrae. This means that the lower lumbar and sacral roots are 20–25 mm long in adult animals, making them easy to isolate on the bath electrodes.

The size of the animal from which the tissue is taken will be determined by the aspect of spinal activity to be investigated. Experience has shown that the dorsal horn of the hamster spinal cord is active at all weights and ages, up to very old animals weighing 150–160 g. The ventral horn seems to be more sensitive to hypoxia. Good motor reflexes can be evoked from the lumbar cord in animals up to 30 g body weight, but above this size the responses are reduced, and can seldom be seen in animals above 50 g. Hamsters in this weight range will be between 3–6 weeks of age and are best described as 'juveniles'; they are weaned and fully mobile, but not yet sexually mature and only weigh about one third of the body weight of an old animal.

It is a good idea when first developing this preparation to start work on animals weighing 20–30 g. Cords from these animals should show good activity in both the dorsal and ventral roots. This will establish that the composition of the bathing medium and the conditions in the recording chamber are satisfactory, before moving on to tissue from older animals where viability may not be so good.

3. Dissection

3.1 Equipment

To perform the dissection the following equipment will be required:

- large scalpel or postmortem knife for decapitation (e.g. Swann Morton PM40)
- scalpel, e.g. Swann Morton No. 22

The spinal cord as an in vitro preparation

- large pointed scissors (6 cm blades)
- medium dissecting scissors for removing surplus tissue (3 cm blades)
- large dissecting forceps
- 2 pairs watchmakers' forceps, size 4 or 5
- fine spring-loaded scissors for cutting bone
- fine spring-loaded scissors for cutting soft tissue
- dissecting chamber (see *Figure 1*)
- anaesthetic chamber
- dissecting microscope

Two pairs of fine scissors will be required because once one of them has been used to cut the bone of the vertebrae they will be too blunt to use on soft tissue. It is a good idea to mark the bone-cutting pair clearly, so that you do not accidentally ruin both pairs.

Before starting the dissection, ensure that everything is ready. The check list in *Protocol 1* shows the items that ought to be prepared. *Protocol 2* provides a guide to the stages in the dissection.

Protocol 1. Procedures to carry out before dissection

Attention to detail is extremely important and can make the difference between the success and failure of this preparation. The following check list is designed to highlight items that should be prepared *before* starting the dissection.

1. aCSF—ensure that sufficient has been prepared (at least 4 litres), and that it has been well chilled. Keep surplus on ice for use in the dissection. See *Table 1* for details of the composition of the aCSF.

2. Circulation—set up the circulation through the experimental chamber *before* starting the dissection. Measure the flow rate by collecting the fluid returning to the reservoir into a measuring cylinder. If it is less than 20 ml/min, find out what is wrong and correct it. If necessary, dismantle the whole system and clean it out—there will not be time if you wait until the cord is ready for mounting in the recording chamber.

3. Gasses—start gassing all of the reservoirs that are going to be used in the experiment with 95% O_2/5% CO_2 before commencing the dissection. This will ensure that the solutions are saturated when the cord is placed in the chamber. Is there sufficient gas remaining in the cylinder to last for the duration of the experiment? If not, replace the cylinder or ensure that a spare is readily available.

4. Calcium—ensure that the correct quantity of calcium has been added to the circulating solution, and that it has not precipitated.

5. Temperature—make sure that the temperature control on the water bath is operating correctly, and that the cord will not be cooked when you place it in the recording chamber. Overheating is often indicated by an excessive number of gas bubbles adhering to the base of the recording chamber as the oxygen comes out of solution.

6. Dissection equipment—ensure that it is all ready and laid out in a manner to make everything easy to find. The early stages of the dissection need to be executed rapidly. There will not be time to change the scalpel blade or search for a missing pair of scissors.

3.2 Anaesthetic

It is possible to decapitate small hamsters without anaesthetic, but it is not advisable to attempt this with animals larger than 20 g body weight. In the UK, you should seek the advice of your Home Office Inspector; the procedures require Home Office approval by licence. Unless the animals are exceptionally tame, the absence of a long tail, the presence of much loose skin around the neck, and an excitable nature combined with a formidable set of incisors make manual restraint difficult. It is more humane to anaesthetize the animal prior to killing.

The choice of anaesthetic is largely determined by convenience and user preference. The anaesthetic agent will be washed out of the tissue during the recovery period and is unlikely to affect activity, although it is known that the effects of pentobarbitone and chloralose persist for longer than urethane and halothane. The difficulty in handling sometimes aggressive animals means that the use of a volatile anaesthetic such as halothane or methoxyflurane in a sealed chamber is often preferred. Detailed instructions on the handling and use of anaesthetics with hamsters are given in reference 4.

Protocol 2. Procedures in the dissection

1. Check that the equipment and perfusion system are ready (*Protocol 1*).
2. Anaesthetize animal.
3. Decapitate animal.
4. Rapidly remove a block of tissue containing the vertebral column.
5. Chill the vertebral column in *cold* aCSF on ice for at least 10 min.
6. Clean muscle and fat off the ventral surface of the vertebrae.
7. Remove the bone of the vertebrae from the *ventral* surface.
8. Identify and mark root L1 (See Section 5.1.1).
9. Carefully excise the cord and remove it from the vertebral canal.

Protocol 2. *Continued*

10. Remove the dura and separate the roots.
11. Hemisect or strip the cord if necessary.
12. Mount the cord in the recording chamber using tungsten pins (*Protocol 3*).

3.3 Removal of the vertebral column

Having anaesthetized the animal, sever the head at the base of the skull using the postmortem knife. Expose the muscles of the back by making a longitudinal incision through the skin along the midline of the back from the neck to the base of the tail. Holding the tail with the large forceps, insert one blade of the large scissors into the anus, and rapidly cut through the body wall from the anus, through the ribs, to the neck on both sides of the vertebral column. It is important to avoid accidentally cutting across the vertebral column at this stage. Ideally, leave about 1 cm of tissue on either side of the vertebral column, which will help with handling the tissue during the rest of the dissection. The freed vertebral column is then plunged into a beaker containing approximately 200 ml of chilled aCSF (*Table 1*) on ice to cool it as rapidly as possible.

This is the only part of the dissection that requires speed. With practice it should be possible to kill the animal and get the vertebral column into the cooling solution in about 30 sec. Thorough rapid cooling is important to preserve the viability of the tissue by minimizing the effects of hypoxia. Ensure that the temperature of the cooling solution is below 4 °C *before* you start the dissection. This is best done by keeping the solution in a refrigerator for at least 8 h before you start. It is not good enough to show the solution to the ice whilst you are inducing anaesthesia. This will result in deserved failure.

The block of tissue containing the vertebral column should be left to cool for about 10 min before being transferred to the dissection chamber. Any adhering viscera should be removed at this stage using medium-size dissecting scissors before the tissue is pinned to the base of the dissecting chamber, ventral surface uppermost, under a layer of chilled aCSF.

Throughout the rest of the dissection it is important that the temperature is kept as low as possible, certainly below 10 °C. This is achieved by maintaining a flow of chilled aCSF through the bath, which also has the effect of improving the visibility of the tissue by washing away any blood and fat droplets that may be released during the dissection. Provided the temperature is kept low, it is not necessary to oxygenate the bathing medium. In skilled hands the cord dissection should take no more than 20 min, but dissections lasting up to 2 h have been performed without noticeable deterioration of the preparation.

Table 1. Composition of aCSF

	Concentration (mM)[a]
NaCl	118
KCl	3.0
NaHCO$_3$	24
MgSO$_4$	1.0
Glucose	12
CaCl$_2$	2.5

[a] All solutions are made up with deionized distilled water

The composition of aCSF is shown in *Table 1*. A 10 × concentrated stock solution is made of all the components *except* the calcium chloride and kept refrigerated until required. All solutions are made up with *deionized distilled water*, and the stock solution is diluted with this when needed. Following dilution, the solution is gassed with 95% O$_2$/5% CO$_2$ for at least 10 min to bring the pH to 7.4 before the calcium chloride is added, usually in the form of a 100 mM solution. If the solution turns milky on addition of the calcium chloride then the calcium has precipitated, probably because the pH is too high, and the solution should not be used.

As a rough guide each preparation usually uses 2–3 litres of solution during the dissection, and 1 litre in the circulating system (2 × 500-ml reservoirs). The day before the proposed experiment 4 litres of aCSF are made up, *without the calcium*, and kept in the refrigerator overnight. On the morning of the experiment 1 litre is placed in the perfusion system, gassed with 95% O$_2$/5% CO$_2$, and the calcium added. The rest is used for the dissection, without gassing or the addition of calcium.

3.4 Dissection of the cord

3.4.1 Cleaning the vertebrae

The rest of the dissection is best performed under the dissecting microscope. With the vertebral column pinned to the base of the dissection chamber (*Figure 1*), ventral surface uppermost, use the fine dissecting scissors to remove as much of the fat and muscle overlaying the ventral surface of the vertebrae as possible. Removal of the vertebrae is much easier if there is a minimum of muscle and tendon adhering to them. If it is the intention to dissect out peripheral nerves attached to the cord, it is important to take care not to sever the nerves as they exit from the vertebral column during this process. Particular care needs to be taken in the lumbar region where the nerves pass through a thick layer of muscle overlying the vertebrae.

3.4.2 Bone removal

Having cleaned the vertebrae, the next stage is to expose the spinal cord by removing the ventral surface of the vertebrae along the length of the cord.

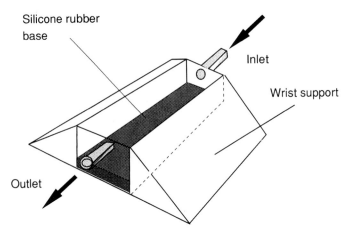

Figure 1. Diagram of the dissecting chamber.

(a) Working from the neck end, grasp the first vertebra with the watchmakers' forceps and lift it slightly so that one blade of the bone-cutting scissors can be inserted between the cord and the bone, with the other blade on the outside of the vertebra, and cut through the bone.

(b) In this way, make two cuts on the ventro-lateral 'corners' of the spinal canal to free the first lamina of bone. This should be peeled upwards, but left attached to the rest of the vertebrae by its caudal edge, providing a useful handle.

(c) Pulling upwards on the freed lamina will allow the bone cutters to be inserted under the second segment and the process repeated. It is important to keep the point of the blade as close to the inner surface of the vertebral column as possible to avoid damage to underlying structures by jabbing it into the cord.

(d) In this way it is possible to remove the ventral laminae from the cervical, thoracic, and lumbar regions without damaging the underlying cord.

(e) In the sacral region, the spinal canal becomes narrow, and the bones are relatively thick, making dissection difficult. In the hamster, the spinal cord does not extend this far, and it is only the sacral roots that are liable to be damaged.

It takes some practice to cut the vertebrae in the best position; if the cut is too ventral then the opening will be too narrow and it will be difficult to free the cord. If the cuts are too lateral there is a great danger of damaging the roots where they exit between the vertebrae. In the thoracic region, the ideal position for the cuts in the bone are close to the point at which the ribs articulate with the vertebrae. In the lumbar region, there are no such guides and the best position can only be learnt through experience.

3.4.3 Excision of the cord

Having exposed the full length of the cord, the next stage is to free it from the vertebral canal.

(a) Again, working from the rostral end, grasp the membranes around the cord firmly using watchmakers' forceps. In younger animals, the dura may be too delicate to sustain much tension, and it may be necessary to hold the cord by one of the spinal roots, or even by the tissue of the cervical cord itself.

(b) Lift the cord slightly so that the blades of the fine scissors can be slipped between the cord and the bone of the vertebrae to cut the roots as close as possible to their point of exit from the spinal canal.

(c) Repeat the process on both sides of the cord for each segment, gradually lifting the cord free from the bone, cutting through any membrane holding the cord in place as you go.

It should be your aim to remove the cord complete with the dorsal and ventral roots still joined at the dorsal root ganglia. At the sacral end it will be necessary to cut across the few remaining roots where they disappear into vertebral canal at the caudal limit of the laminectomy. This will completely free the cord which should be pinned to the base of the dissection chamber using one or two fine tungsten pins (see *Protocol 3*). Fine tungsten pins and needles can be produced by etching the end of lengths of tungsten wire in fused potassium nitrite. Excellent pins and needles can be made from lengths of straightened wire.

Protocol 3. Manufacture of tungsten pins and needles

Materials

- 0.25 mm diameter tungsten wire which is commercially available (Clark Electromedical, UK—Cat No. TW10–3)
- *wear goggles, gloves, and protective clothing when working with molten KNO_2, and work in a shielded environment such as a fume cupboard: KNO_2 can splash out of the container and is a strong oxidizing agent*

Method

1. Heat a small amount (i.e. a 5 ml teaspoonful) of KNO_2 crystals in a shallow metal tin such as a tobacco tin, over a bunsen burner. Do not overfill the tin; the molten KNO_2 should not cover the base of the tin. The etching process occurs best in the thin layer at the meniscus of the fluid with the bare metal bottom of the tin. Place the gas flame under this area and adjust the heat until the metal glows dull red.

The spinal cord as an in vitro preparation

Protocol 3. *Continued*

2. Holding a length of tungsten wire in pliers or haemostats, dip 2–3 mm of the end into the KNO_2 at at angle of approximately 30° to the base of the tin. The immersed tungsten will start to effervesce. Drag the tip of the wire out of the KNO_2 onto the bare metal. As the tungsten passes through the meniscus there will be a vigorous reaction and the tip of the wire will become red hot and be rapidly etched, often accompanied by sparking.
3. After about 5 sec the tip should be etched to a fine point. Remove the wire from the heat and wash the tip in water to remove adhering KNO_2.
4. Cut the pin/needle to the required length using wire cutters. *Care should be taken when cutting the pins. The tungsten is very hard and the pins can fly off at unexpected angles.* It is best to insert the sharp end in a block of expanded polystyrene before cutting. This is also a convenient way of storing the pins.

3.4.4 Dorsal and ventral roots

The next stage of the dissection is to free the roots, which in the lumbar and sacral regions are tightly wrapped in a layer of dura mater, and to separate the dorsal and ventral roots. Using tungsten pins, firmly secure the rostral end of the cord to the base of the bath with the dorsal surface uppermost (see Section 5.1.1). Use another pin through the dura of the *cauda equina* (the bundle of roots in the sacral region) to stretch the cord and hold it straight. With watchmakers' forceps grasp the dura close to the dorsal root ganglion on one of the larger lumbar roots and cut a small hole in the membrane. Insert one blade of the scissors into this hole and cut the dura along the midline for the length of the cord. This will allow the bundle of roots to unwrap and each root can then be individually cut free from the adhering dura.

As each pair of roots is freed from the dura, the dorsal and ventral roots should be separated by cutting through the ventral root, which at this stage will be the root that emerges from the undersurface of the cord, at the point where it joins the dorsal root, just proximal to the dorsal root ganglion. This will leave the dorsal root ganglion attached to the dorsal root, which can be a valuable aid to identification of the dorsal and ventral roots when the cord is placed in the recording chamber.

Free all the sacral and lumbar roots on both sides of the cord, and as many of the thoracic roots as are needed. The lumbar roots tend to stick together and can be gently separated by running the point of a fine tungsten needle between them. Avoid damaging the roots by bending them at sharp angles, or by pulling too hard, especially at the point where they enter the cord.

Finally, turn the cord over and remove as much as possible of the dura from the ventral surface. The cord can now be transferred to the recording chamber if it is not to be hemisected or stripped of pia.

3.5 Hemisection

With hemisection the aim is to obtain a good a midline section with minimum damage to the cord. There are two ways of achieving this:

(a) *Tearing*. Tearing can give a very clean midline split of the cord, and is best used on short cords taken from younger animals where the connective tissue is not too strong. Using two pairs of watchmakers' forceps, the rostral end of the cord is grasped and the right and left halves pulled apart using a steady lateral pull. Considerable damage is done to the rostral end of the cord by the forceps, but the rest of the cord will split very cleanly. If the connective tissue connecting the roots from the right and left sides has not been completely removed, there is a great danger of stripping the roots off one side of the cord.

(b) *Cutting*. An alternative way to hemisect cords is to cut down the midline using a knife made from a fragment of razor blade mounted on a holder (see knife preparation, *Protocol 4*). This is best done from the ventral surface where the groove down the midline provides a convenient knife guide. It is important that both halves of the cord are pinned securely at the rostral end, and that the cord is held taut to minimize the tendency to bend as the cut progresses. The process is made easier and more accurate if two cuts are made along the length of the cord, the first using just the tip of the blade to cut through the pia covering the midline. In the lumbar region, the same effect can often be achieved by pulling off the central blood vessel using fine forceps. The second cut goes through the depth of the cord, penetrating into the base of the bath. The aim should be to make one smooth movement cutting cleanly from rostral to caudal straight down the midline. In practice this can be difficult and the cut often ends with damage to the tracts in the midline, or with one half of the cord having the dorsal tracts corresponding to both sides of the cord.

When hemisecting the cord using the cutting technique, it is also important to ensure that none of the roots are trapped under the cord or cross the midline, because they will be cut off if they are not repositioned to their correct side.

Protocol 4. Manufacture of small knives

Small, sharp blades suitable for hemisecting the cord can be made from razor blades.

1. Use blades made from non-stainless steel, which are reasonably thick and can be broken cleanly. The razor blades sold for shaving are usually too thin, but the single-sided blades that are available from many suppliers are suitable.

Protocol 4. *Continued*

2. Hold the blade firmly in the jaws of a stout pair of pliers and snap off a triangular piece from one corner using another pair of pliers (*Figure 2*). Ideally, the section snapped off should be approximately one-third of the length of the blade. With a little practice it is possible to produce two sharp-pointed knives from each single-sided razor blade.
3. Mount the blade fragment in the jaws of a small pin chuck with handle, or glue it with an epoxy resin adhesive to the flattened end of a glass rod to form a handle.
4. Use a fresh blade for each dissection—being made of mild steel the blades lose their edge rapidly and are prone to rusting.

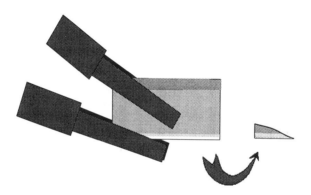

Figure 2. Method of grasping a single-sided razor blade in the jaws of two pairs of pliers to break off a fragment for use as a fine knife.

3.6 Stripping

In some cases where an intact cord is needed but the roots are not required, it can be an advantage to strip off the pia surrounding the cord to reduce the diffusion barrier to oxygen and other chemicals entering the tissue. This is a very simple procedure. The cord is pinned firmly to the base of the dissecting chamber by the rostral end, and the rostral end of the pia is grasped with a pair of watchmakers' forceps and rolled back, rather like skinning a sausage. As the pia folds back the roots will be stripped off, leaving a bare cord. At the beginning of the lumbar expansion it will be necessary to make a longitudinal slit in the pia to allow it to roll over the bulge, before it can be pulled free from the taper at the sacral end of the cord.

4. The perfusion system

4.1 Circulation

The system used to ensure a good perfusion of the experimental chamber is shown in *Figure 3*. Artificial CSF is stored in two or three reservoirs made from 500 ml plastic measuring cylinders, the outlets of which are connected to the feed tube of the heat exchanger using a system of three-way taps. This allows switching between the different reservoirs with minimum disturbance to the flow. The solutions in the reservoirs are all continuously gassed with 95% O_2/5% CO_2.

The perfusion fluid flows through a heat exchanger to adjust its temperature just prior to entering the experimental chamber. The heat exchanger is constructed of a straight 40 cm length of stainless steel tube through which the aCSF flows. This is surrounded by a 2 cm diameter plastic tube to form a water jacket. Water from a thermostatically controlled water bath is pumped through the water jacket to adjust the temperature of the aCSF. The heat exchanger is placed as close to the inlet of the experimental bath as possible to reduce changes in the temperature of the aCSF before it enters the bath.

The aCSF flows through the recording chamber and is *sucked* out of the small well through a syringe needle connected to a peristaltic pump, then is returned to the reservoir. It is important to avoid 'tidal' flow of the solution in the bath caused by variations in the fluid level at the outlet. To prevent this, the suction tube should be terminated by a narrow-diameter tube such as a syringe needle, and the pump should be set to a high speed so that there is an approximately 50:50 aCSF/air mixture sucked through the needle. This will

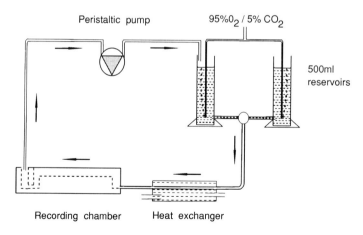

Figure 3. Diagram of the system for recirculating aCSF.

also allow accurate control of the bath fluid level by adjusting the height of the outlet needle.

4.2 The recording chamber

The recording chamber (*Figure 4*) consists of a central slot 5 mm wide milled in a 7 × 12 cm block of Perspex. A black silicone rubber compound is used to form a floor in the slot so that pins can be inserted to secure the cord. The perfusion fluid flows over the cord and is sucked out of a small circular well at the end of the chamber. Pairs of silver wires (0.8 mm diameter) are inserted into holes drilled at intervals along the side of the bath to form the stimulation/recording electrodes. Better isolation from the bathing medium is obtained if these are raised 1 mm above the level of the surrounding bath on a platform of Perspex. The bath is raised on 1 cm high pillars, and connections are made to the ends of the silver wires where they project below the bath.

Earthing is provided by a strip of silver ribbon which is glued to the inner edge of the central recording chamber so that it is immersed in the perfusion fluid.

4.3 Flow rate

It is of vital importance that an adequate flow rate be maintained through the experimental chamber. This is the only means by which oxygen can be delivered to the tissue. When working at 27 °C a flow rate of *at least* 20 ml/min

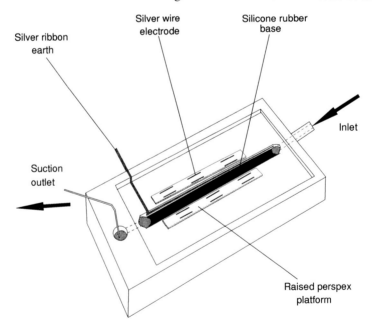

Figure 4. Diagram of the recording chamber.

is required, and at higher temperatures, or where there is a particularly large bulk of tissue in the recording chamber the flow rate is best increased to 30 or 40 ml/min.

The flow into the bath is gravity-fed and is determined by the head of aCSF in the reservoirs and the diameter of the tubes on the inlet side of the system. In practice, the diameter of the tubes is usually fixed, being limited by the diameter of the orifices through the three-way taps, and flow rate is usually adjusted by changing the height of the reservoirs. Anything that reduces the diameter of the tubing, such as the growth of microorganisms, or circulating debris, can drastically reduce the flow rate through the system with disastrous consequences. It is therefore important to keep all the tubes and taps clean.

4.4 Cleaning

At the end of each experiment, the system should be flushed through with distilled water to remove as much organic material as possible. Once a week, or more frequently if there is noticeable slowing of the flow rate through the circulation, the tubing and heat exchanger should be dismantled and thoroughly cleaned by pushing short lengths of pipe cleaners through the tubing to scour out any deposits. An occasional flush through with dilute hydrochloric acid will help to remove any calcium precipitates that may have formed on the inside of the tubing. If the plastic tubing is badly contaminated, the best strategy is to replace it.

5. Recording

5.1 Mounting and recovery

Following dissection, the cord is transferred to the recording chamber. If it has to be moved any distance, it can be supported on a piece of filter paper moistened with aCSF. Use a tungsten pin inserted through the rostral end of the cord both for handling the tissue, and to secure it to the base of the recording chamber. The rostral end of the tissue should be positioned near the inlet to the recording chamber so that the fluid flows from rostral to caudal over the preparation and does not bend the roots back upon themselves. Ensure that the whole length of the cord is under the surface of the solution. During the course of an experiment, bubbles of gas will often collect under the cord and tend to lift it to the surface. To prevent this, place two or three more pins along the length of the cord to secure the thoracic and lumbar regions. Where possible, insert the pins through adhering connective tissue or dura but, if necessary, tungsten pins can be inserted directly into the cord itself, although they will cause some local disruption and may damage longitudinal fibre tracts.

The second half of an hemisected preparation can be kept for up to 8 h in a beaker of oxygenated aCSF for later use, providing the temperature is kept at or below 4 °C. In our experience, tissue kept in this way for longer periods seldom shows satisfactory electrical activity.

When mounting the cord in the recording chamber, it is worth giving some thought to the recording arrangements that you are going to use to avoid unnecessary handling of the tissue. If recordings are to be made from the dorsal horn or dorsal roots of the whole cord, it is usually best to mount it with the dorsal surface uppermost. Motoneurones are probably best approached from the ventral surface. See Section 5.1.1 for the identification of the dorsal and ventral aspects of the cord.

When using a hemisected preparation, the layout of the tissue is best visualized from the cut midline of the preparation; electrode penetration also is easier from this aspect, there being no pia to impede penetration. The hemisected cord is, therefore, best mounted with the lateral surface of the cord facing the floor of the chamber.

Once mounted in the recording chamber, the cord requires 2–3 h incubation at room temperature (18–20 °C) before full activity is obtained. Recordings from the dorsal roots made during this time often show initial activity consisting of two or three units firing continuously at relatively high frequencies (10–50 Hz). This activity seldom persists for more than 10 min after mounting a root on the electrodes and is usually ascribed to damage to the roots caused by handling. The cord is often quiescent for 30–60 min before bursts of spontaneous action potentials can be detected emerging from the cord along the dorsal roots. The frequency of this activity will increase during the next 1–2 h before settling down to a steady level which will be maintained for the subsequent 8–10 h.

It is useful to monitor the spontaneous dorsal root activity, even if it is not to be directly the object of your investigation, because experience has shown that it provides a sensitive index of the state of the cord. In a healthy cord the dorsal roots will fire bursts of 5–10 action potentials with 1–2 sec between bursts (*Figure 5a*). Such cords will usually show strong segmental reflexes and good longitudinal conduction. Less healthy or damaged cords have a reduced rate of spontaneous activity, often tending to show continuous firing of one or two units with few bursts of action potentials. In the early stages of hypoxia, as occurs when the gas supply is interrupted, there is a crescendo of continuous unit activity in the dorsal roots as the oxygen supply diminishes, before activity abruptly ceases. This can be a very valuable early warning of impending disaster; corrective action at this stage will usually save the cord.

5.1.1 Identification of parts of the cord

Once the cord has been dissected free and removed from the landmarks provided by the vertebrae, it can be difficult to identify the different parts of the cord if you do not have previous experience.

i. *Dorsal or ventral—whole cord*

It is important to be sure if you are stimulating or recording from the dorsal or

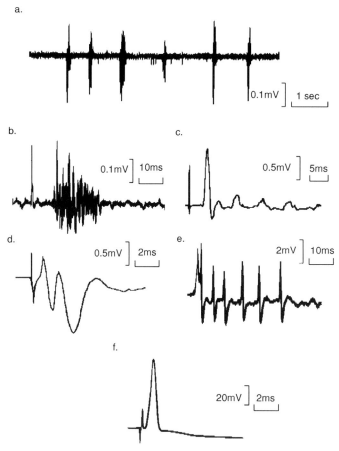

Figure 5. Some examples of recordings made from isolated hamster spinal cords. (a) Bursts of spontaneous action potentials in a lumbar dorsal root. (b) Dorsal root reflex evoked in a lumbar dorsal root by electrical stimulation of an adjacent dorsal root. (c) Ventral root reflex recorded from a lumbar ventral root following stimulation of the corresponding lumbar dorsal root. (d) Field potential in the lumbar dorsal horn evoked by stimulation of a lumbar dorsal root. (e) Action potentials from a single unit in the lumbar dorsal horn recorded with a microelectrode following stimulation of a lumbar dorsal root. (f) Intracellular recording made from a motoneurone in the lumbar cord antidromically activated by stimulation of a ventral root.

ventral aspects of the cord. There are a number of clues that can be used as a guide:

(a) Shape—the dorsal cord has a rounded, convex appearance, whereas the ventral aspect is either flat or slightly depressed along the midline.
(b) Colour—the ventral surface of the cord is a uniform white colour. The dorsal cord has a white stripe down the midline which marks the dorsal tracts. The stripe is flanked laterally on both sides by an area of pink

colouration where the well vascularized dorsal horn comes very close to the surface of the cord.

(c) Blood vessels—in the midline of the dorsal surface of the lumbar cord there is a large prominent vein which is usually filled with red blood. There are also blood vessels on the ventral surface of the lumbar cord but these are smaller and less prominent.

ii. Dorsal or ventral—hemisected cord

With the cut midline surface upwards in the recording chamber, the dorsal and ventral white tracts can be clearly distinguished from the central strip of pink-coloured grey matter in all but the youngest cords where myelination is very incomplete. In the lumbar region, it is possible to identify dorsal and ventral surfaces by the shape of the white matter because the diameter of the cord reduces as it approaches the sacral region. The ventral tracts maintain a constant width relative to the overall diameter of the cord and taper down smoothly. The dorsal tracts taper sharply over the last few lumbar segments and form a very narrow band in the sacral region.

iii. Identification of roots

Once the landmarks of the vertebrae are removed, it is very difficult to positively identify individual roots. This problem is best addressed by identifying and labelling as a marker a particular root during the dissection, from which other roots can be identified by counting their offset. One of the more convenient roots to use as a marker is L1. This gives rise to a nerve which passes over the inner dorsal surface of the abdominal wall and is easy to dissect intact with the spinal roots. It is the first spinal nerve to leave the vertebral column that is not associated with a rib. A tie placed around this nerve still attached to the cord makes an excellent marker.

5.2 Recording techniques

All the standard extracellular and intracellular recording techniques can be used on isolated mammalian spinal cord. To date, we have concentrated mainly on extracellular recording from the roots using silver wire electrodes, or on field potentials and unitary activity using glass micropipettes. Standard intracellular recordings are also possible, and intrafibre recordings have been made from afferent fibres in the dorsal horn.

5.2.1 Root recording

Recordings can be made from the spinal roots, or from filaments dissected from them, by placing the root over one of the pairs of silver wire electrodes mounted on the sides of the bath, and making connection to a suitable AC amplifier. The largest problem with this technique is the danger of desiccation. The roots are very susceptible to water loss, and exposure to a dry atmosphere for only 5 min is sufficient to kill them. When dried out, they change from their normal opaque white colour to become very thin, transparent, and stiff. It is useless to attempt to record from roots in this condition.

Desiccation is prevented by covering the roots, once they are in place over the electrodes, by a layer of grease made from petroleum jelly (Vaseline) mixed with sufficient liquid paraffin to make it soft enough to allow it to be squeezed out of a 2 ml syringe with a 19-gauge needle. It is important to ensure that the entire surface of the root is covered with grease, taking it right down to the water line in the bath to avoid desiccation at the point of exit of the root from the bathing fluid.

Another common problem is trapping a pocket of air that is in contact with the atmosphere under the grease. This can cause the gradual desiccation of the root over 1–2 h and is best avoided by placing a thin layer of grease in the gap between the two wire electrodes before the root is draped over them. Care, however, must be taken to ensure that the grease does not get between the electrodes and the roots and prevent good recordings from being made. Contact of the roots with the electrodes can often be improved by using the point of a tungsten needle to scrape the connective tissue along the edge of the root against the wire of the electrode.

Using this system of recording, action potentials of 200–500 µV can be recorded against baseline noise of 20–40 µV (*Figure 5a,b*).

5.2.2 Microelectrode recording

Recordings can be made of single unit activity in the isolated spinal cord preparations using conventional glass micropipettes, which are filled with 2 M NaCl and have a resistance of 10–20 MΩ. In intact cords, it will be necessary to make a small tear in the pia with tungsten needles to allow the electrode to penetrate. In the hemisected cord, penetration can be made from the cut medial surface, avoiding the pial connective tissue.

Field potentials evoked by stimulation are best recorded using blunt (< 5 MΩ) glass micropipettes filled with 1 M NaCl. Such large-diameter electrodes give excellent low noise recordings, and field potentials of 1–2 mV are not exceptional (*Figure 5d*). There will, however, be considerable leakage of electrolyte into the tissue which can cause the signal to decline if the electrode is left in the same spot for more than 10 min. This can be overcome by plugging the tip of the electrode with 1% agar jelly made up in 1 M NaCl, to prevent leakage and allow stable potentials to be recorded from the same spot for many hours.

5.2.3 Other techniques

Two other recording techniques that can be used for recording from the spinal roots are suction electrode (see Chapter 12) and sucrose-gap, although the author does not have much direct experience of them.

Suction electrodes are made from blunt, flame-polished glass pipettes into

The spinal cord as an in vitro preparation

which the root is sucked by negative pressure. A chloride-coated silver wire is used to record potentials from the fluid surrounding the root in the pipette relative to a reference electrode in the bathing medium surrounding the cord. This system has the advantage that it can be used to record from short roots and nerves that are not long enough to reach the wire electrodes on the side of the bath. DC potentials recorded from the roots reflect changes in the polarization of the nerve terminals (dorsal roots) or motoneurones (ventral roots) within the cord. The main disadvantage is the practical problem of ensuring a tight fit of the electrode around the nerve to enable good recordings to be made. Suction electrodes can also be used for stimulation.

In the sucrose gap technique, the nerve or root is stretched between two recording chambers filled with saline. The chambers are separated by a third chamber which is filled with a non-conducting solution of isotonic sucrose. Changes in potential in the recording chamber closest to the spinal cord are recorded relative to the reference potential in the distant chamber using a DC amplifier, and reflect potential changes occurring within the cord.

5.3 Stimulation

Electrical stimuli can be delivered to the dorsal roots using the silver wire electrodes mounted along the sides of the bath. Because the roots are well isolated from the bath and are covered in grease, maximal stimulation can usually be achieved at low stimulus intensities. A stimulus of 10–15 V with a duration of 0.1 msec is usually found to be supramaximal for the large-diameter fibres in the lumbar dorsal roots, producing large evoked potentials in the dorsal horn and reflex responses from the ventral roots. The bipolar configuration also helps to reduce the size of the stimulus artefact.

Where it is necessary to excite the cord itself, stimulation electrodes can be made of pairs of fine Teflon-insulated stainless steel or silver wires twisted together and cut off at the tip so that only the bare end of the metal is exposed. These can be mounted on micromanipulators and placed on the surface of the tissue to produce local stimulation of the longitudinal tracts. Because of the greater electrical resistance of these electrodes, and the fact that stimulation is taking place under the saline, larger stimulus voltages are usually needed than when using the electrodes mounted on the sides of the bath. Stimuli of 20–50 V, 0.1 msec duration, are often needed to produce effective stimulation.

5.4 Example recordings/manipulations

Some examples of typical recordings made from whole hamster spinal cord preparations are shown in *Figure 5*.

(a) *Spontaneous dorsal root activity*. This should be easy to record, consisting of bursts of 5–20 action potentials at intervals of approximately 1 sec, which can be recorded from any of the dorsal roots using silver wire electrodes (*Figure 5a*). Strong spontaneous activity is usually a sign of good synaptic activity and reflexes in the dorsal horn of the preparation. Continuous or random firing is only normally seen in cords from animals less than 4 weeks old. In older cords it is usually a sign of distress, caused by either damage during dissection or hypoxia.

(b) *Evoked dorsal root reflexes.* Electrical stimulation of one of the dorsal roots evokes a burst of action potentials emerging antidromically from the adjacent dorsal roots—the dorsal root reflex (*Figure 5b*). In roots immediately adjacent to the stimulated root, the dorsal root reflex will be seen after a delay fo 8–12 msec and lasts for approximately 150–200 msec. In more distant ipsilateral roots, and in contralateral roots, the delay will be longer. Both spontaneous dorsal root activity and the evoked dorsal root reflex are temperature sensitive, with an optimum between 25–27 °C, declining sharply above this temperature (*Figure 6a*).

(c) *Spontaneous ventral root activity.* When maintained in magnesium-free medium, at a temperature between 20 and 25 °C, cords from juvenile animals display spontaneous activity in the lumbar ventral roots. Unlike the spontaneous dorsal root activity, this usually consists of randomly firing single action potentials.

(d) *Evoked ventral root potentials.* In cords taken from younger animals, stimulation of the dorsal roots, or the ascending branches of the afferent fibres in the dorsal tracts, evokes a reflex discharge in the motoneurones

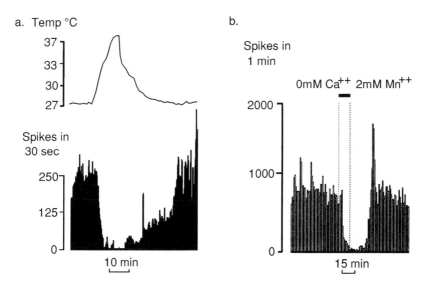

Figure 6. (a) The effects of increasing the temperature (upper graph) upon the frequency of spontaneous dorsal root activity (lower plot). (b) Blockade of spontaneous dorsal root activity by replacement of the calcium in the bathing medium with manganese.

which can be recorded from the ventral roots as a volley of action potentials leaving the cord (*Figure 5c*). In good preparations, this consists of the two classical components—a synchronous, short latency monosynaptic potential, which may be up to 5 mV in amplitude, followed by an asynchronous polysynaptic discharge. In less healthy preparations, the monosynaptic component is often reduced or absent, but the polysynaptic component may still be present.

(e) *Field potentials.* Stimulation of the dorsal roots evokes field potentials which can be detected by blunt (< 5 MΩ) microelectrodes within the cord (*Figure 5d*). Good ventral horn responses will only be obtained in cords from younger animals, but strong dorsal horn field potentials should be detected in cords of all ages. These may be up to 2 mV in amplitude (depending on the recording site) and often show at least three components:

  i. short latency potential of the incoming volley in the stimulated afferent fibres

  ii. intermediate latency (5–10 msec) wave as synaptic activation excites postsynaptic cells

  iii. late, slow potential which reaches a peak 50–100 msec after stimulation and persists for up to 500 msec. This potential will only be properly recorded if DC recording is used; AC recording will reduce the size of the potential, and tend to make it biphasic

(f) *Unit recording.* Using sharp micropipettes (resistance > 10 MΩ) both extracellular and intracellular recordings can be made from individual units within the cord. Some examples of extracellular recordings are shown in *Figure 5e* and of an intracellular recording in *Figure 5f*.

(g) *Synaptic blockade.* One of the major advantages of isolated CNS preparations is the ability to manipulate the composition of the medium bathing the tissue. It is often useful to be able to block synaptic activity in the preparation to determine the nature of a recorded potential (i.e. synaptic or non-synaptic). This can be done by replacing the calcium in the bathing medium with manganese (*Figure 6b*). Depending upon the thickness of the tissue and the amount of surrounding connective tissue, all synaptic activity will be blocked in 10–20 min. On return to normal aCSF the synaptic activity returns.

6. Temperature

With such a large block of tissue the limitations imposed by the diffusion of oxygen into it mean that changes in temperature can have dramatic effects upon its responses (*Figure 6a*). It is important to monitor continuously the bath temperature to ensure that it is maintained at the optimum for the response under examination. In addition, changes in the temperature can provide valuable early warning of problems. A decrease in the flow rate will usually be accompanied by a reduction in the temperature as the solution has longer to cool before reaching the bath. Changes in the temperature when switching between reservoirs can also cause large changes in the activity of the tissue; care must be taken to ensure that this does not interfere with the results obtained.

Experience has shown that in the whole-cord preparation, dorsal root and dorsal horn activity reaches a peak at 25–27 °C and falls off with temperatures above and below this. At 30–32 °C both the spontaneous dorsal root activity and the evoked dorsal root reflex are suppressed (*Figure 6a*), to recover on reduction of the temperature again. This effect is unlikely to be entirely due to hypoxia because in *in vivo* preparations the dorsal root reflex is best seen when the temperature of the cord is reduced, and disappears at temperatures close to body temperature.

Ventral root activity is more sensitive to temperature. Evoked ventral root reflexes are seldom seen at temperatures above 25 °C and are optimum around 20–22 °C. If the temperature does rise much above this, then the response will be reduced, usually irreversibly. In regions of the world that have high temperatures (and in Britain occasionally), it may be necessary to reduce the temperature of the bath by passing the inlet tube through an ice slurry.

7. Troubleshooting

7.1 Possible problems and remedies

One of the main problems faced by the experimenter when attempting to set up an isolated spinal cord system for the first time is to identify what is wrong if no activity can be recorded from the preparation. There is not really any substitute for the presence of an expert who is familiar with the preparation and with what might go wrong. The following check list will help to narrow down the possibilities, although it should be remembered that there may be more than one fault.

(a) *Is the recording apparatus working?* This is best tested by deliberately introducing mains noise into the recording and observing it on an oscilloscope, or listening to it on a loudspeaker. If recordings are being made from the silver wires on the side of the bath, then touch each wire in turn with the tips of a pair of uninsulated watchmakers' forceps. If the recording system is working, this should cause a large sine wave at mains frequency (50 Hz in the UK, 60 Hz in USA) to appear on the oscilloscope and a humming noise to be heard on the loudspeaker. When using microelectrode recording, a similar effect can be achieved by bringing your hand close to the electrode.

(b) *Is the gas supply adequate?* A careful eye needs to be kept on the gas supply to the reservoirs, especially during the recovery period, when the tendency is to leave the preparation for a few hours to do something important, like sitting on a committee or having lunch. If the gas supply has been interrupted for less than 20 min and the temperature has not yet increased above 20 °C, there is a good chance of obtaining a reasonable recovery if a vigorous gas flow is restored as soon as possible—it is worth a try.

(c) *Is there adequate calcium in the solution?* A low concentration of calcium in the aCSF (< 2 mM) will cause synaptic responses to be reduced or absent, although conducted responses in stimulated axons will persist and may be enhanced. This can be caused by either forgetting to add the calcium to the aCSF or if the calcium has precipitated from the solution. Precipitation can usually be seen under the dissecting microscope as flocculent white particles in the recording bath, and is often caused by a high pH in the solution because of inadequate gassing with 5% CO_2. In either case, replace the aCSF with fresh solution containing the correct calcium concentration; synaptic activity should return in about 30 min.

(d) *Is the flow rate through the bath too slow?* The supply of oxygen to the tissue is entirely dependent upon an adequate flow of aCSF through the recording chamber. This should be at least 20 ml/min, and preferably closer to 30 ml/min. It is important to check the flow rate as part of the setup procedure, and again during the experiment if there is any suggestion of problems. Blockages at narrow points of the system, such as in three-way taps, may be caused by floating debris and must be cleared as soon as possible.

(e) *Is the temperature in the recording chamber correct?* A high temperature can kill the preparation, especially the ventral horn. A low temperature might indicate a drop in flow rate and consequent hypoxia.

(f) *Is the preparation totally dead?* Different parts of the preparation die at different rates, and this can be helpful in determining the source of the problem. Ventral horn responses and the ventral root reflex evoked following stimulation of the dorsal roots are the most sensitive components. If these are missing, but there are still dorsal horn and dorsal root responses to stimulation, this suggests that the oxygen supply to the ventral horn is inadequate, due either to working at too high a temperature, too low a flow rate through the bath, or using animals that are too large. If dorsal root/horn activity is also absent, it suggests either that most synapses are not functioning well or the preparation is totally dead. To test for this, try direct recording from a dorsal root and use a mobile stimulating electrode to excite either the base of the root where it enters the cord, or the dorsal tracts just rostral to the point of entry of the root into the cord. In either case, it should be possible to evoke a large, short-latency antidromic action potential in the root. If this cannot be obtained, then the preparation is totally dead and it indicates a fundamental problem. If there is a good antidromic action potential but no evoked response, it implies that the failure is occurring at the level of the synapses and it should be possible to improve the situation by:

  i. increasing the calcium level
  ii. improved oxygenation
  iii. changing the temperature
  iv. as a last resort adding a low concentration of 4-aminopyridine (5 µM) to the bathing medium to improve synaptic activity

7.2 If the preparation is completely dead

If you cannot get even antidromic conducted activity then something is fundamentally wrong.

- Check that you have made up the aCSF properly; were your calculations correct?
- If someone else in the laboratory is successfully working on isolated mammalian CNS preparations, try some of their aCSF or ask them to try some of yours on their preparation (at the end of their experiment in case it kills it!)
- Check that nothing went wrong during the dissection. Did it take more than a minute to get the vertebral column into the cold aCSF? Was the temperature in the dissecting bath kept below 10 °C?
- Could any toxic substance have crept in at any stage, e.g. are the dissecting instruments also being used on fixed tissue, or is histology being performed in the vicinity?

If you cannot sort it out the first time keep trying. It generally takes new recruits to our laboratories between 10 and 20 preparations to obtain good activity, but thereafter we usually achieve a success rate in keeping the cord alive of better than 80%.

References

1. Kerkut, G. A. and Wheal, H. V. (ed.) (1981). *Electrophysiology of Isolated Mammalian CNS Preparations*. Academic Press, London.
2. Dingledine, R. (ed.) (1984). *Brain Slices*. Plenum Press, New York.
3. Schurr, A., Teyler, T. J., and Tseng, M. T. (ed.). *Brain Slices: Fundamentals, Applications and Implications*. Karger, Basel.
4. Green, C. J. (1979). *Animal Anaesthesia*. Laboratory Animals Ltd., London.

10

Mathematical description of neuronal firing patterns

J. L. HINDMARSH and R. M. ROSE

1. Introduction

The purpose of this chapter is to explain in a simple and direct manner how to build and develop mathematical models of the behaviour of the membrane potential of a neurone. We do this by presenting models of increasing sophistication. In Section 2, a model due to FitzHugh is used to illustrate the main ideas. This is followed in Section 3 by a series of models, each an elaboration of the previous one. The models of these two sections are qualitative in nature and independent of the precise details of the various ionic currents present in the cell.

Almost all the ideas we introduce below are represented graphically. The necessary graphs can be easily reproduced using a microcomputer, and the reader is strongly encouraged to do so. Some notes on the BASIC programs that have been used to produce *Figures 6a* and *6b* are to be found in the appendix to this chapter.

2. Theory

2.1 Prediction of the time course of the membrane potential

Figure 1 shows the graph of the membrane potential x against time t, for a repetitively firing neuron, which has been computed using the model to be described in Section 3.1.

This graph sufficiently resembles recordings reproduced elsewhere for us to use it to explain the following problem of prediction. Given that the membrane potential at time t_0 is x_0, can we predict the membrane potential at some later time?

To make this question more precise, suppose we consider the times $t_n = nt_s$, $n = 0, 1, 2, \ldots$, where t_s is a fixed time step, and let the membrane potential at time t_n, $x(t_n)$, be x_n. Can we find a rule, or formula, f that gives $x_{n+1} = f(x_n)$? For example, suppose f were defined by the rule

$$f(x) = x + t_s(x^3 - 3x^2),$$

Mathematical description of neuronal firing patterns

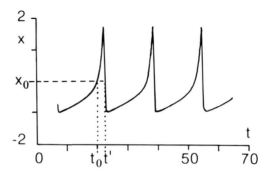

Figure 1. The numerical solution of Equations 6 with $l = 0.5$ and $t_s = 0.05$, x has the value x_0 at times t_0 and t', but the subsequent values are different (see Section 2.1).

then this rule gives the $(n + 1)$th value of the membrane potential x_{n+1}, in terms of the nth value of the membrane potential x_n, as:

$$x_{n+1} = f(x_n) = x_n + t_s (x_n^3 - 3x_n^2)$$

where this equation holds for $n = 0, 1, 2, \ldots$

The answer to the above question is no. Such a simple rule cannot exist because the membrane potential is the same at times t_0 and t' of *Figure 1* but the subsequent potentials are different; they increase after t_0 and decrease after t'.

We take this difference to mean that the state of the cell at time t' differs from the state of the cell at time t_0. To allow for this difference in state, we introduce an additional variable y. This variable may, for example, represent the conductivity of an ion channel. We will simply call y an internal variable.

The state of the cell at time t_n is now described by the pair of numbers (x_n, y_n), and we need a rule to give the state of the cell (x_{n+1}, y_{n+1}) at time t_{n+1} in terms of the state of the cell at time t_n. For states specified by two variables, the sort of rules we will look at will take the form:

$$x_{n+1} = x_n + t_s P(x_n, y_n)$$
$$y_{n+1} = y_n + t_s Q(x_n, y_n) \quad (1)$$

where P and Q denote functions of (or formulae in) two variables (see for example Equations 2 below). Note that these equations give both the value of the membrane potential and the internal variable at time $t_{n+1} = t_n + t_s$ in terms of their values at time t_n for all values of integer $n \geq 0$.

2.2 The FitzHugh model

A particularly simple example of a rule of the form specified by Equations 1 is given by:

J. L. Hindmarsh and R. M. Rose

$$x_{n+1} = x_n + t_s\,(y_n - x_n^3 + 3x_n^2)$$

$$y_{n+1} = y_n + t_s\,(1 - 4x_n - y_n) \tag{2}$$

This rule comes from equations used by FitzHugh.

As we will see below, they predict behaviour which is in many respects similar to that of a real cell. We will call a predictive rule, such as Equations 2, designed to simulate the behaviour of a real cell, a (mathematical) model. *Protocol 1* shows how Equations 2 may be used to compute future states of the cell.

Protocol 1. Computing future states using Equations 2

1. Assume, for example, that the state at time $t_0 = 0$, $(x_0, y_0) = (-1, 0)$ and take the time step $t_s = 0.05$.
2. Put $x_0 = -1$, $y_0 = 0$, and $t_s = 0.05$ into the right-hand side of Equations 2 (with $n = 0$), which gives $x_1 = -0.8$, $y_1 = 0.25$. Thus, the state at time $t_1 = 0.05$ is $(-0.8, 0.25)$.
3. Use these values for (x_1, y_1) in Equations 2 (with $n = 1$) to find the state at time $t_2 = 0.1$. It is $(-0.6659, 0.4475)$.
4. Consult *Table 1* for the first five states computed in this way. Although this calculation can be done with a pocket calculator, it is ideally suited for a computer.
5. Consult Section 4 for some notes on BASIC programs used to calculate these states.

In *Figures 2a* and *2b* we show the points (t_n, x_n) and (t_n, y_n) for $n = 0$ to 200 as dots. The first five of these correspond to the values in *Table 1*. By 'joining the dots' we obtain the graph of the membrane potential against time. In all subsequent figures we will join the dots and draw these graphs as continuous curves.

Since our models are qualitative, rather than quantitative, we are not using

Table 1. Future states calculated as in *Protocol 1*

Time	x	y
0	−1	0
0.05	−0.8	0.25
0.1	−0.6659	0.4475
0.15	−0.562247816	0.608305
0.2	−0.475527213	0.740339313
0.25	−0.399214871	0.84842779

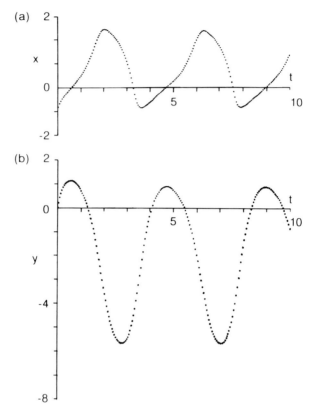

Figure 2. Plots of (a) x_n and (b) y_n against time for Equations 2. The system was started at $x_0 = -1$ and $y_0 = 0$; and successive values of x_n and y_n are shown as dots for $n = 0, 1, 2, \ldots 200$ with $t_s = 0.05$.

physical units. For this reason we do not indicate millivolts or milliseconds, etc. on the figures.

2.3 The 'state space' and 'state diagram'

In addition to plotting the points (t_n, x_n) and (t_n, y_n), we can also plot the points (x_n, y_n) in what is called the 'state space' (also called the phase space). For example, *Figure 3a* shows the states of *Table 1* followed by the next 195 states. The path formed by these states is called the 'state path' (also called the phase path). Different starting states will give rise to different state paths as shown in *Figure 3b*. The picture provided by *Figure 3b* is called a 'state diagram' (also called a phase diagram).

One reason why we use this state space is to provide another picture of how the state changes in time. A more important reason is that some features of the state diagram can be seen directly from the Equations 1 without computing state paths.

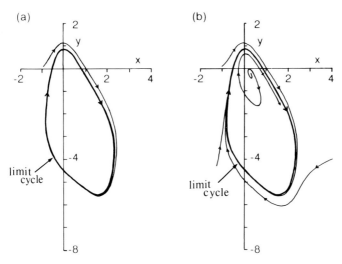

Figure 3. Phase diagrams for Equations 2 with $t_s = 0.05$. (a) For one starting point $(x_0 = -1, y_0 = 0)$, (b) For four starting points: $(x_0 = -1, y_0 = 0)$, $(x_0 = -2, y_0 = -5)$, $(x_0 = 4, y = -4)$, and $(x_0 = 0.3, y_0 = -2)$. Note the appearance of a limit cycle in both cases.

Since these features are reflected in features of the time course of the membrane potential, we have an important link between the equations and the membrane potential. For example, suppose we change the equations by adding a term to represent an external current. We may want to see how this term effects the time course of the membrane potential. We will discuss this in Section 2.8. We will see in Section 3 how changes in the equations defining a model will alter the way the model behaves.

2.4 The 'tangent vector'

Suppose we have a model whose equations are of the form of Equations 1. The relation between the state (x_{n+1}, y_{n+1}) and (x_n, y_n) may be expressed graphically as in *Figure 4*.

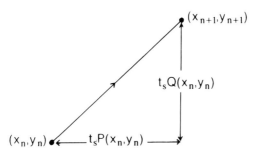

Figure 4. Representation of target vector $(P(x_n, y_n), Q(x_n, y_n))$ at x_n, y_n. See text for further explanation.

Making the timestep t_s smaller brings (x_{n+1}, y_{n+1}) closer to (x_n, y_n), but does not change its direction from (x_n, y_n). This direction is fixed by the two equations $P(x_n, y_n)$ and $Q(x_n, y_n)$.

If, for example, $P(x_n, y_n) > 0$ then $x_{n+1} > x_n$; in other words, if $P(x_n, y_n) > 0$, then the membrane potential increases from the state (x_n, y_n) to the state (x_{n+1}, y_{n+1}). If, for example, $Q(x_n, y_n) = 0$ then $y_{n+1} = y_n$ and the value of the internal variable does not change from the state (x_n, y_n) to the state (x_{n+1}, y_{n+1}).

We call the pair of numbers $(P(x, y), Q(x, y))$ the 'tangent vector' at (x, y). In the case of the FitzHugh model defined by Equations 2, the tangent vector for the state (x, y) is $(y - x^3 + 3x^2, 1 - 4x - y)$. So for the state $(1, 2)$ the tangent vector is $(4, -5)$, which tells us that from this state the membrane potential will increase $(4 > 0)$ and the internal variable will decrease $(-5 < 0)$.

We can now say that, once the functions P and Q in Equations 1 are specified, these equations tell us what the tangent vector is for any state of the system. This in turn tells us the direction of the next state. The distance away from the next state depends on both the magnitude of t_s and the magnitudes of $P(x, y)$ and $Q(x, y)$.

Although we have defined the tangent vector at (x, y) as the pair of numbers $(P(x, y), Q(x, y))$, we may think of it as the geometric tangent vector to the state path through (x, y).

Finally, we should note that Equations 1 can be rewritten in the form:

$$\frac{x_{n+1} - x_n}{t_s} = P(x_n, y_n)$$

$$\frac{y_{n+1} - y_n}{t_s} = Q(x_n, y_n) \qquad (3)$$

The first of these says that the change in membrane potential in time t_s is given by $P(x_n, y_n)$. In other words, the time rate of change of the membrane potential, when the system is in state (x_n, y_n), is $P(x_n, y_n)$. Similarly, the second equation says that the time rate of change of the internal variable, when the system is in state (x_n, y_n), is given by $Q(x_n, y_n)$.

2.5 The 'nullclines'

The state $(1, -2)$ has, for the FitzHugh model given by Equations 2, the tangent vector $(0, 1)$. This means that the membrane potential will not change from this state to the next state. In geometric terms the tangent vector is vertical.

We can locate the states where the tangent vector is vertical by finding all

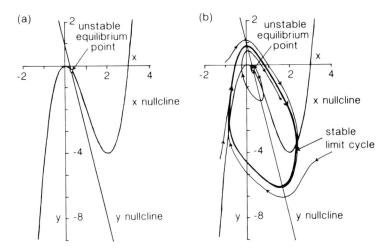

Figure 5. State diagram for Equations 4 with $I = 1$ (or equivalently Equations 2, $t_s = 0.05$. (a) The nullclines and unstable equilibrium point. (b) Here the nullclines in (a) are added to the phase diagram of *Figure 3(b)* to give the 'state diagram'. See Section 2.8.

states (x, y) for which $P(x, y) = 0$. For the FitzHugh model the co-ordinates of these states must satisfy the equation $y - x^3 + 3x^2 = 0$, i.e. they must lie on the curve $y = x^3 - 3x^2$ in the state space. This curve, shown in *Figure 5a*, is called the *x nullcline*.

Similarly, the states where the tangent vector is horizontal, the *y nullcline*, must have co-ordinates for which $Q(x, y) = 0$. For the FitzHugh model these states lie on the line $y = 1 - 4x$, also shown in *Figure 5a*.

The next figure (*Figure 5b*) combines the state diagram of *Figure 3a* with the nullclines shown in *Figure 5a*. Note that the state paths cross the x nullclines vertically and the y nullclines horizontally.

2.6 The 'equilibrium points'

A state of special interest is a state that belongs to both the x nullcline and y nullcline; a state that is a point of intersection of these curves, e.g. the state (0.317 . . . , −2.706 . . .) of the FitzHugh model. The tangent vector for such a state is (0, 0) and so this state does not change. It is in equilibrium, and called an *equilibrium point*. This state path stating at an equilibrium point does not go anywhere and simply consists of that one point. The membrane potential for a system, whose state is an equilibrium point, will be constant and its time course a horizontal line.

What is now of interest is what happens to states near an equilibrium point. For example, the state diagram for the FitzHugh model (*Figure 3b*) suggests that state paths starting near the equilibrium point (0.317 . . . , −2.706 . . .) spiral away from it. In fact, this is the case for all paths starting near this equilibrium point. This means that, although this model predicts that the state of the system can remain fixed with the constant membrane potential 0.317 ... internal variable −2.706 ..., a slight disturbance may change this state to a nearby state that does not remain constant. As time passes, the state changes along a spiral path in the state space that takes it well away from the equilibrium point. Equilibrium points, with this property that a slight disturbance may produce a large change in the state of the system, are called unstable equilibrium points.

We will see later that we can also have equilibrium points which have the property that, when disturbed to a nearby state, this state changes along a state path that returns it closer and closer to the equilibrium point. Such states are called stable equilibrium points.

It is clearly of interest to be able to determine whether or not an equilibrium point is stable. One way to do this is to use the equations of the model to see what happens to nearby states. Since this is determined by the tangent vectors it is perhaps not surprising that it is possible to decide the matter directly from the form of the functions P and Q. For details of this procedure, see reference 1.

2.7 The 'limit cycle'

The most significant feature of the state diagram (*Figure 3a*) of the FitzHugh model is that all state paths (except the state path consisting of the one state that is the equilibrium point) approach a state path that forms a closed curve. If the system starts in a state on this closed curve, its state will return to its starting state after a certain time T and again after time $2T$, etc.; the state of the system will go through a periodic change. In other words, no matter what the initial state of the system (ignoring the unstable equilibrium point), the system will eventually settle into a regular periodic rhythm. Furthermore, even if it is disturbed from this rhythm is will settle back into it. The periodic behaviour in the state space is reflected in the periodic behaviour of the time courses of the membrane potential and the internal variable as shown in *Figures 2a* and *2b*.

A closed state path as described above is called a 'stable limit cycle'. Its stability, like that of the equilibrium point, lies in the fact that, if at some time the state of the system is disturbed from a state on the limit cycle to a nearby state, then the state path from this nearby state will gradually get closer and closer to the limit cycle.

2.8 The effect of varying a parameter

We now change Equations 2 to:

$$X_{n+1} = x_n + t_s (y - x^3 + 3x^2 - 1 + I)$$

$$Y_{n+1} = y_n + t_s (1 - 4x_n - y_n) \tag{4}$$

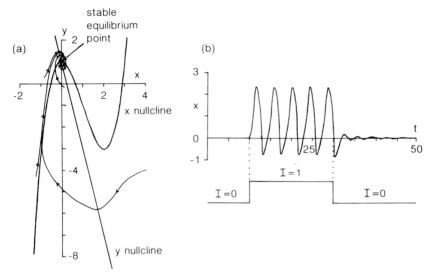

Figure 6. The effect of varying the external current, I. (a) 'State diagram' for Equations (4) with $I = 0$. Phase paths converge to a stable 'equilibrium point' as $x = 0$. (b) The numerical solution of Equations 4 for a step change in I. When $I = 0$ the phase diagram is as shown in (a). When the value of I is changed to 1 the phase diagram is as shown in *Figure 5b* and the 'equilibrium point' becomes unstable. This results in repetitive firing as shown here. Computer programs for this Figure are given in Section 4.

where I is a parameter, intended to represent the effect of an external current applied to the cell.

Clearly, the tangent vector for the state (x, y) depends on the value of I. This means that the location of the nullclines and of the equilibrium point will also depend on the value of I. Indeed the whole state diagram will depend on the value I.

If we put $I = 1$ in Equations 4, then they become the same as Equations 2 and their state diagram and nullclines are as shown in *Figure 5b*. If we put $I = 0$, then the state diagram and nullclines are as shown in *Figure 6a*.

The effect of the change in value of I has been to lift the x nullcline upwards by 1 unit. In consequence the equilibrium point, the point of intersection of the x and y nullclines, has also moved. The most striking changes are that the equilibrium point has now become stable, in that nearby paths are approaching it, and the stable limit cycle has disappeared.

2.9 The effect of an external current step

Now let us use Equations 4 to make a model of a cell that has an external current step applied to it.

At time $t = 0$ we assume that the state of the cell is $(0, 1)$ and use Equations 4, with $I = 0$, to predict the values of the membrane potential and the internal

Mathematical description of neuronal firing patterns

variable at times $t_n = nt_s$ for $t_s = 0.05$ and $n = 0$ to 200. Since we are using a time step $t_s = 0.05$, this will give us the state of the cell for times between 0 and 10. *Figure 6b* shows the time course of the membrane potential. As we expect from the state diagram for $I = 0$ (*Figure 6a*), the membrane potential is constant because the initial state is the stable equilibrium point.

At time 10 we change the value of I from 0 to 1. In order to predict the subsequent change in state, we now use Equations 4 with $I = 1$. The relevant state diagram is now that of *Figure 5b*, so we expect to see periodic behaviour. This is the case as shown in *Figure 6b*. At time 30 the value of I is changed back to 0 and the membrane potential returns to its equilibrium value.

In order to gain experience, we suggest the following 'numerical experiment'. Run the calculation above just as described until $t = 10$. Change the value of I to 1 and again run the calculation as above, except that the value of I should be changed back to 0 after a very short time, say at $t = 10.5$.

2.10 Differential equation models

Almost any attempt to obtain further information about the Fitzhugh model will result in finding it presented as a pair of differential equations. For example the differential equations

$$\frac{dx}{dt} = y - x^3 + 3x^2 - 1 + I$$

$$\frac{dy}{dt} = 1 - 4x - y \tag{5}$$

are one version of the FitzHugh model. These equations give the time rates of change of the state variables x and y in terms of these variables. To solve these differential equations is to find two functions of t, $x(t)$ and $y(t)$, such that at all times t:

$$\frac{dx(t)}{dt} = y(t) - x(t)^3 + 3x(t) - 1 + I$$

$$\frac{dy(t)}{dt} = 1 - 4x(t) - y(t)$$

Once we have found $x(t)$ and $y(t)$, we can plot the time courses of x and y and the state path in the same way as above.

To produce a formula for $x(t)$ (such as $x(t) = 2 \sin t - \cos t$) is usually impossible, and these solutions have to be found using numerical methods.

The simplest method of solution of these equations is Euler's method which says that if $x(t)$ and $y(t)$ are the values of x and y at time t, then $x(t + t_s)$, the value of x at time $t + t_s$, is given by:

$$x(t + t_s) \approx x(t) + t_s \frac{dx}{dt}$$

In words, the value of the membrane potential at time $t + t_s$ is its value at time t plus the product of the time step with the time rate of change of the membrane potential, as defined by Equations 5 for the state of the system at time t. Using Equations 5 this becomes:

$$x(t + t_s) \approx x(t) + t_s(y(t) - x(t)^3 + 3x(t)^2 - 1 + I)$$

which, if we put $x_{n+1} = x(t + t_s)$, $x_n = x(t)$, and $y_n = y(t)$, is the same as the first of Equations 4. Similarly,

$$y(t + t_s) \approx y(t) + t_s (1 - 4x(t) - y(t))$$

What this means is that state paths for the differential equations:

$$\frac{dx}{dt} = P(x, y)$$

$$\frac{dy}{dt} = Q(x, y)$$

may be obtained from the equations:

$$x_{n+1} = x_n + t_s P(x_n, y_n)$$

$$y_{n+1} = y_n + t_s Q(x_n, y_n)$$

albeit approximately. How closely the state paths, computed by Euler's method, approximate the solution of the differential equations depends on the size of t_s. The smaller the step size t_s, the more accurate the approximation.

2.11 Choice of step size

The time courses of membrane potential x and internal variable y, or the state paths, that we obtain from equations such as Equations 4 will depend on the magnitude of t_s. This may readily be seen by trying the values of 0.01, 0.1, and 0.5, instead of the value 0.05 used above. Since the results we obtain depend on t_s, we must specify a value in order to define the model properly.

Mathematical description of neuronal firing patterns

The following simple rule may be used to find an appropriate value for t_s. Start with a value so large that taking a smaller value gives a significantly different time course or state path. Continue to take smaller values for t_s until doing so makes no significant change. This final value of t_s is an appropriate choice.

Of course, the smaller the value of t_s, the more steps, and so more computations are needed to find the time courses, or state paths, over a fixed interval of time.

3. The mathematical models

3.1 The repetitively firing neurone

The first model in this section results from a very small modification to the FitzHugh model. The defining equations are:

$$x_{n+1} = x_n + t_s(y_n - x_n^3 + 3x_n^2 - 1 + I)$$

$$y_{n+1} = y_n + t_s(1 - 4x_n^2 - y_n) \qquad (6)$$

The only change is that the term $4x_n$ has become $4x_n^2$. This means that the y nullcline is no longer a straight line but is now parabolic, and, in a region of state space of interest, lies close to the x nullcline. The state diagram and nullclines are shown in *Figure 7a* for the case where $I = 1$. We see a state path that is stable limit cycle and that part of this path passes between the two nullclines. If this path is plotted on a microcomputer screen, it can be seen that the state changes quite slowly on the limit cycle as it passes between the nullclines and comparatively quickly elsewhere. The reason for this is as follows. States on, for example, the x nullcline have (see Section 2) zero time rate of change of x and, similarly, states *near* the x nullcline will have a *small* time rate of change of x. Similarly, states near the y nullcline will have a small time rate of change of y. Consequently, states near both nullclines will be states for which both x and y are changing slowly in time. Plotting the time course of x shows that this slow change corresponds to a slow interspike recovery period (*Protocol 2*).

Protocol 2. Calculation of frequency–current relationship for Equations 6

1. Use values of I in the range 0.5–2.5; start the system in the state $(-1, 0)$ say.
2. Plot the time course of x. This should be periodic as shown in *Figure 1*, where $I = 0.5$, and *Figure 7b*, where $I = 1$.

3. For each value of I, measure the frequency of the resulting periodic motion.
4. Finally, plot a graph of the frequency against current.

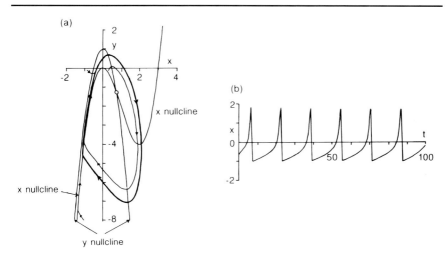

Figure 7. The repetitive firing neurone. (a) The 'state diagram' and (b) the plot of x against time for Equations 6 with $I = 1$. Note close proximity of x and y nullclines for $x < 0$ in (a). This accounts for the long interspike interval in (b) (see Section 3.1).

3.2 Triggered firing

This model is a simple modification of the previous one, where the x and y nullclines were in close proximity in part of the state space. Because of this closeness it requires only a small change in the equations of the model to make the nullclines intersect again, thus creating new equilibrium points (EP) for the model. This happens for the model:

$$x_{n+1} = x_n + t_s(y_n - x_n^3 + 3x_n^2 + I)$$
$$y_{n+1} = y_n + t_s(1 - 5x_n^2 - y_n) \qquad (7)$$

Note that Equations 7 are not quite the same as Equations 6. The term $-4x_n^2$ has become $-5x_n^2$.

We will consider this model for two different values of the parameter I.

3.2.1 The case where $I = 0.125$

Here the x and y nullclines are the curves: $y = x^3 - 3x^2 - 0.125$ and $y = 1 - 5x^2$, respectively. The points of intersection of these curves, the EPs, are the points whose x co-ordinate satisfies the equation:

$$x^3 - 3x^2 - 0.125 = 1 - 5x^2$$

These points are $(-1.5, -10.25)$, $(-1.151\ldots, -5.627\ldots)$, and $(0.651\ldots, -1{,}121\ldots)$.

An investigation of the state diagram will reveal that, depending on where they begin, the state paths either approach the point $(-1.5, -10.25)$, which is a stable EP, or they approach a stable limit cycle, as shown in *Figure 8a*. This means that for $I = 0.125$ the model has two stable states, a silent resting state and an active firing state.

3.2.2 The case where $I = 1$

The x and y nullclines intersect once at a point which is an unstable EP. All 'state paths' approach a 'stable limit cycle'. The model has no resting state.

The difference between these two cases is that as the value of I is increased from 0.125 to 1, the x nullcline, which is the curve $y = x^3 - 3x^2 - I$, is lowered by 0.875. Since the position of the y nullcline does not depend on I, it remains fixed as I varies. As the x nullcline moves down, its points of intersection with the y nullcline move towards each other, meet when $I = 5/27$, and disappear.

Consideration of the two cases discussed above suggests that it may be possible to start in the stable EP of the model, with $I = 0.125$, and apply a current step increasing the value of I to 1, which switches the model from its stable EP into a 'stable limit cycle'. The important point is that, after the current step is switched off, the model remains in its firing mode.

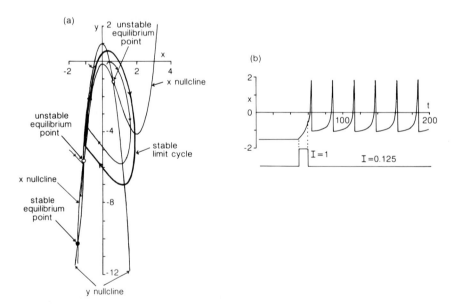

Figure 8. Triggered repetitive firing. (a) The 'state diagram' for Equations 7 with $I = 0.125$. (b) Here the system starts at $(x_0 = -1.5, y_0 = -10.25, I = 0.125)$ corresponding to the left-most stable equilibrium point in (a). Application of a current step of $I = 1$ switches the system to the limit cycle in (a) corresponding to repetitive firing in (b) (see Section 3.2).

Investigation of this possibility is the purpose of the next numerical experiment (*Protocol 3*).

Protocol 3. Determining the effect of an external current step on the model defined by Equations 7
1. Follow the procedure described in Section 2.9.
2. Start the model in the state $(-1.5, -10.25)$ with $I = 0.125$ and find the state at time 50.
3. Put $I = 1$ and see how this state changes from time 50 to time 60.
4. Finally return to $I = 0.125$, which should leave the model in its stable limit cycle as shown in *Figure 8b*.

The current step in the above experiment is shorter than the time taken to complete one cycle of the limit cycle, and is more appropriately called a current pulse. This situation where a model is switched from resting to repetitive firing by a short current pulse is called triggered firing.

Protocol 4. Further experiment with the model defined by Equations 7 with $I = 0.125$
1. Start the model in a state whose state path leads to the stable limit cycle.
2. Let the model settle into steady repetitive firing.
3. Now find a suitable negative step that can be applied to the model that will switch it from repetitive firing to its resting state.

3.3 Adaptation

In the previous model of triggered firing, the model, when firing, would do so indefinitely. We now want to introduce a further elaboration that will cause this firing to cease and return the cell to its resting state. This is done by introducing a third state variable z, called the 'adaptation variable', and defining a new model by the equations:

$$x_{n+1} = x_n + t_s(y_n - x_n^3 + 3x_n^2 + I)$$
$$y_{n+1} = y_n + t_s(1 - 5x_n^2 - y_n)$$
$$z_{n+1} = z_n + t_s\, r(s(x_{n+1}) - z_n) \quad (8)$$

where r and s are positive contants whose values will be discussed below. The first two of these equations are the same as those of Equations 7. The third equation, giving the rule for the change in value of z, says that z will change by an amount $\delta z = t_s\, r\, (s(x_{n+1}) - z_n)$.

If $z_n < s(x_{n+1})$ then $\delta z > 0$ and z will increase. Conversely, if $z_n > s(x_{n+1})$ then z will decrease. This means that the value of z changes, at each time step, towards the value of $s(x_{n+1})$.

Suppose we take the model defined by Equations 7 with $I = 0.125$, $r = 0.005$, and $s = 2$, and start in the state where $x = -1.5$, $y = -10.25$, and $z = -0.75$. Since $(-1.5, -10.25)$ is a stable EP, for the model defined by the first two of Equations 8, the x and y variables will not change; however, the z variable will change towards -1, which is the value of $s(x_{n+1})$, since $s = 2$ and $x_n = -1.5$ at each step (*Protocol 5*). The change in z at each step will be quite small because $r = 0.005$ is small.

Protocol 5. Finding the time course of the adaptation variable z(t)

1. Take the model defined by Equations 8 with $I = 0.125$, $r = 0.005$, and $s = 2$.
2. Start in the state $(-1.5, -10.25, -0.75)$.
3. Find the time course of $z(t)$.
4. Repeat steps 1–3, but this time apply a current pulse by putting $I = 1$ from 200 to time 210.[a]
5. Find the time course of $z(t)$.
6. Compare the result with that shown in *Figure 9*.

[a] The effect of this will be to make the model defined by the first two of Equations 8 switch to repetitive firing. This will mean the value of x_n will change from its resting value of -1.5 to values varying periodically between -1 and -2, approximately. This in turn means that the value of z will change (slowly) towards the value of $s(x_n + 1)$, which is now varying between 0 and 6.

It will be observed that in *Figure 9* the value of $z(t)$ approximates the time average value of $s(x_{n+1})$, which is nearer to 0 than 6 because the membrane potential spends more time in the recovery phase than in the action potential range. We have seen above that the value of z can be changed by the change of state, as determined by the first two of Equations 8. We now complete the circle of influence by including z_n in the first of Equations 8 to get the model:

$$x_{n+1} = x_n + t_s\, (y_n - x_n^3 + 3x_n^2 + I - z_n)$$

$$y_{n+1} = y_n + t_s\, (1 - 5x_n^2 - y_n)$$

$$z_{n+1} = z_n + t_s\, (r\, (s(x_n + 1.5) - z_n)) \tag{9}$$

Notice that the equation for z_{n+1} has been altered so that z is changing towards $s(x_n + 1.5)$. This means that if $x_n = -1.5$ then z is changing towards 0. The behaviour of these equations is investigated in *Protocol 6*.

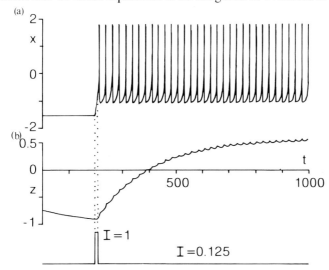

Figure 9. Triggered repetitive firing. Plot of (a) x against time, (b) z against time for Equations 8 with $r = 0.005$ and $s = 2$. The initial state is $(-1.5, -10.25, -0.75)$. The application of a current pulse of $I = 1$ switched the system into repetitive firing as in *Figure 8*. Note that the slow change in z has no effect on the time course of x (uncoupled case).

Protocol 6. Termination of burst by adaptation

1. Take the model defined by Equations 9 with $I = 0.125$, $r = 0.0007$, and $s = 0.5$.
2. Start in the state $(-1.5, -10.25, 0)$.
3. Find the time course of $x(t)$ and $z(t)$.
4. Note that these graphs should be horizontal lines because the state $(-2.5, -10.25, 0)$ is a stable equilibrium point for the model.
5. Verify by observing the behaviour of state paths starting close to this point.
6. Secondly, take the model defined by Equations 9 with $I = 0.125$, $r = 0.0002$, and $s = 0.5$.
7. Start in the state $(-1.5, -10.25, 0)$.
8. Apply a current pulse $I = 1$ from time 200 to time 210 to switch the model into repetitive firing.
9. Find the time course of $x(t)$ and of $z(t)$. These should be as shown in *Figure 10*.

Mathematical description of neuronal firing patterns

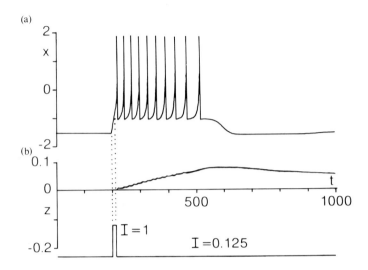

Figure 10. Triggered burst. Plot of (a) x against time, (b) z against time for Equations 9 with r = 0.0002 and s = 0.5. Repetitive firing terminates to give a triggered burst because the z variable is now coupled to the first of Equations 9.

Figure 10 shows that, once the repetitive firing starts, the value of z begins to increase. This means that the value of the term $I - z$ in the first of Equations 9 will decrease. This is equivalent to reducing the value of I. In consequence, the firing rate slows down and the model eventually stops firing. At this points z begins to decrease and the state of the model returns to its stable equilibrium point.

Some insight into this behaviour may be gained from considering the first two of Equations 9. We can regard these as a two-dimensional model whose state diagram depends on the value of $I - z$. As z slowly increases, the x nullcline moves gradually upwards, moving the equilibrium points until the stable limit cycle no longer exists and the state path approaches the stable equilibrium point.

3.4 Burst generation

In Section 3.3 we saw how a third variable could be introduced in order to terminate the triggered firing. This suggests that for the model defined by:

$$x_{n+1} = x_n + t_s (y_n - x_n^3 + 3x_n^2 + I - z_n)$$

$$y_{n+1} = y_n + t_s (1 - 5 x_n^2 - y_n)$$

$$z_{n+1} = z_n + t_s (r(s(x_n + d) - z_n)) \tag{10}$$

where an additional parameter d has been introduced in the third equation, it should be possible to find values of I, r, s, and d for which the model has the following properties:

(a) when the model is firing, z increases until the firing terminates
(b) when the model is not firing, z decreases and becomes sufficiently negative for $I - z$ to be sufficiently positive to start the model firing again

If values of I, r, s, and d can be found, this should result in a model that will alternate between firing and silence, i.e. exhibit bursting. The difficulty, of course, is in finding suitable values. A way of simplifying this problem is as follows.

First, assume that the mechanism required to alternate the model between silence and firing is essentially a subthreshold mechanism. That is, it has to do with the behaviour of the model in the region of state space where the nullclines of the two-dimensional system defined by the first two of Equations 10 are close together. In this region, the state of the model lies close to both nullclines and so, in any state in this region, we will have $y_n \approx 1 - 5x_n^2$.

Second, we replace y_n in the first of Equations 10 by this approximate value. This leaves us with the two equations:

$$x_{n+1} = x_n + t_s(-x_n^3 - 2x_n^2 + 1 + I - z_n)$$
$$z_{n+1} = z_n + t_s(r(s(x_n + d) - z_n)) \quad (11)$$

Now choose values for I, r, s, and d so that this system has a stable limit cycle in which the value of z oscillates between a negative value, which will correspond with firing, and a positive value, which will correspond to silence. In other words, we want the model defined by Equations 11 to have a stable limit cycle.

The x and z nullclines for these equations are given by the curves $z = -x^3 - 2x^2 + 1 + I$ and $z = s(x + d)$, respectively. The first of these curves is a cubic and the second a straight line, as may be seen by plotting their graphs.

Now recall the FitzHugh model of Equations 2 for which the nullclines, shown in *Figure 5a*, also consisted of a cubic curve and a straight line. This model had a stable limit cycle when the straight line intersects the cubic in its middle section. We, therefore, seek values of I, s, and d so that this happens here. The values $I = 0.125$, $s = 2$, and $d = 1$ will achieve this, as shown in *Figure 11a*.

Next choose a value of r. Since this affects the amount by which the z variable changes in each time step, and we have previously required z to change slowly, we take a small value for r, $r = 0.004$. With these values the state diagram does include a limit cycle which is shown in *Figure 11b*.

Having chosen values for I, r, s, and d, return to the full system of Equations 10. Starting in the state $(-1.5, -10, 0)$ with $I = 0.125$, $r = 0.004$,

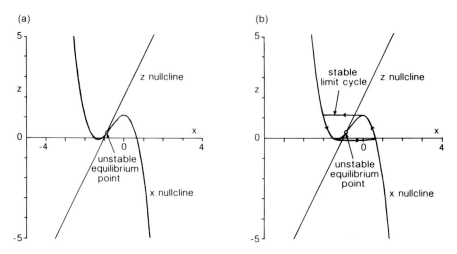

Figure 11. (a) The nullclines of Equations 11 with $I = 0.125$, $s = 2$, and $d = 1$ showing unstable equilibrium point. (b) The nullclines and limit cycle for Equations 11 with $I = 0.125$, $s = 2$, $d = 1$, and $r = 0.004$.

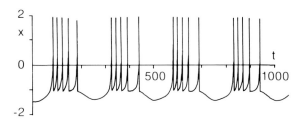

Figure 12. Periodic bursting. The time course of x for Equations 10 starting at $(-1.5, -10, 0)$ with $I = 0.125$, $s = 2$, $d = 1$, and $r = 0.004$.

$s = 2$, and $d = 1$, we obtain the time course of $x(t)$ shown in *Figure 12*. We see that this model shows periodic bursts.

4. Conclusions

The mathematical models discussed in Sections 2 and 3 depended on writing down equations that describe the way the state of the system changes in time. The question arises as to what justification there is for choosing these equations, apart from the fact that their predicted behaviour does resemble that of real neurones. In other words, if we started from a detailed knowledge of the membrane structure, could we deduce equations similar to those we have used? To some extent this is possible, and a most significant step in this direction was taken by Hodgkin and Huxley in 1952. Unfortunately, space does not permit a discussion of this physiological justification of these equations. Instead, we provide a list of references for further reading. For further discussion of state diagrams see reference 1 and for a more advanced discussion of non-linear differential equations see reference 2; more emphasis on the biology is to be found in reference 3. For an introduction to the biophysical approach see reference 4.

References

1. Abraham, R. H. and Shaw, C. D. (1982). *Dynamics: The Geometry of Behaviour Part I. Periodic Behaviour.* Aeriel Press, Santa Cruz.
2. Jordan, D. W. and Smith, P. (1987). *Non-linear Ordinary Differential Equations* (2nd Edition), Clarendon Press, Oxford, (1987).
3. Glass, L. and Mackey, M. C. (1988). *From Clocks to Chaos. The Rhythms of Life.* Princeton University Press.
4. Ferreira, H. G. and Marshall, M. W., (1985). *The Biophysical Basis of Excitability.* Cambridge University Press.

Appendix

Computer programs

As examples of the kind of computer program required to calculate the figures drawn in this chapter, we list overleaf the two programs used to produce *Figures 6a* and *6b*. The programs are aimed at the reader who has little previous knowledge of computing and are written in BASIC, as this is one of the easiest languages to understand. Note that the instructions 'plot' and 'draw axes and graduation marks' are machine dependent, and will depend on the screen co-ordinates used.

Mathematical description of neuronal firing patterns

The program for *Figure 6a* is:

```
10 DEF FN_P(X,Y)=(Y-X*X*X+3*X*X-1+I)
20 DEF FN_Q(X,Y)=(1-4*X-Y)
30 REM Main programme follows
40 I=0:TS=0.05
50 CLS
60 Draw axes and graduation marks
70 REM Draw nullclines
80 FOR X=-5 TO 5 STEP 0.01
90 Y=X*X*X*X-3*X*X+1-I
100 Plot(X,Y)
110 Y=1-4*X
120 Plot (X,Y)
130 NEXT X
140 REM Draw phase paths
150 PROCdrawpath(-1,0,1000)
160 PROCdrawpath(-2,-5,1000)
170 PROCdrawpath(4,-4,1000)
180 PROCdrawpath(0.3,-0.2,1000)
190 DEF PROCdrawpath(X,Y,iter)
200 FOR N=0 TO iter
210    PLOT (X,Y)
220    NEWX=X+TS*FN_P(X,Y)
230    NEWY+Y+TS*FN_Q(X,Y)
240    X=NEWX
250    Y=NEWY
260 NEXT N
270 ENDPROC
280 END
```

The program for *Figure 6b* is:

```
10 DEF FN_P(X,Y,I)=(Y-X*X*X+3*X*X-1+I)
20 DEF FN_Q(X,Y)=(1-4*X-Y)
30 REM Main programme follows
40 I=0:TS=0.05
50 CLS
60 Draw axes and graduation marks
70 REM Plot x against t
80 PROCdrawgraph(0,1,0,200,0)
90 PROCdrawgraph(X,Y,201,600,1)
100 PROCdrawgraph(X,Y,601,1000,0)
110 DEF PROCdrawgraph(X,Y,startiter,enditer,I)
120 FOR N=startiter TO enditer
130    T=N+TS
140    Plot(X,Y)
150    NEWX=X+TS*FN_P(X,Y,I)
160    NEWY=Y+TS*FN_Q(X,Y)
170    X=NEWX
180    Y=NEWY
190 NEXT N
200 ENDPROC
210 END
```

PART IV

Nerve cell pharmacology

11

Iontophoresis in the mammalian central nervous system

M. H. T. ROBERTS and T. GOULD

1. Introduction

The microiontophoretic application of drugs onto neurones of the central nervous system can be one of the more seriously misused of pharmacological techniques. It remains, however, almost the only way of applying relatively few molecules rapidly into the vicinity of central synapses. The technique gets closer to mimicking synaptically released neurotransmitter than any other, but its limitations must be thoroughly understood or very misleading data will emerge. This chapter will outline the strengths of the technique, as well as the methodological constraints, and will give practical examples of how it may sensibly be used. It will also discuss common mistakes which should be avoided.

1.1 Basic principles

An electromotive force will move charged particles within its field. Positively charged ions will migrate to the negatively charged electrode and vice versa. The rate at which the ions move is determined by clearly understood principles described by Faraday's law:

$$R = \frac{In}{zF}$$

where R is the rate of ionic movement, I is the current (amps); n is the transport number of the ion, z is the valency of the ion, and F is the Faraday constant. It follows that the ions move in linear proportion to the current passed. For any current, however, ions with different transport numbers will move at different rates. Consider a simplified situation where a solution contains only two monovalent ions, a cation and an anion. The current will be carried by the movements of both ions. If the transport numbers of both ions are identical (and therefore 0.5), both ions will move at equal rates, although in opposite directions. If, however, one is a small ion and the other a larger ion, it is unlikely that the transport numbers of the two ions will be the same, and they will not move at equal rates. A higher (and possibly very much higher) proportion of the current will be carried by the more mobile ion. Commonly, the biologically active ion of a salt is the less mobile.

Microiontophoresis involves the following procedures: An ionized drug solution is placed into a glass tube which has been melted and pulled into a micropipette (*Figure 1*). The fine tip is lowered into the tissue and an electrical current is caused to flow between the tissue and the drug solution. This current is carried by the migrating cations and anions. With one polarity of current, a stream of cations will emerge from the tip of the electrode. This will change, when the charges are reversed, into a stream of anions. If the tip of the electrode is positioned close to a central neurone, then the neurone can be bathed in cations or anions at will.

This introductory description of the technique is seriously inadequate from a practical viewpoint, however, because several other events will occur that are very relevant but are not described above. The most important of these include:

- the effects of the electrical current itself on the cell
- the diffusion of drug ions from the concentrated solution in the electrode
- the dilution of the biologically active ion in the tip of the electrode when the counter ion is ejected for a long period of time
- the effects of charge developing on the inner surface of the glass pipette

These points, and others, are discussed in more detail in the following sections.

2. Preparation of micropipettes

2.1 Manufacture of blanks

The most commonly used iontophoresis micropipette is multibarrelled and contains one barrel that is used as an electrode for recording extracellularly the action potentials produced by neurones in the close vicinity of the tip. This type of micropipette and its manufacture is described in *Protocol 1*. Five-barrelled micropipette blanks are made from glass tubes purchased from Clark Electromedical Instruments. We use two types of tubing, both of which contain fine glass filaments. Recordings are made from one barrel, made of wider diameter tubing than the others, which are used for the iontophoretic ejection of drugs.

Protocol 1. Preparation of multibarrelled micropipette blanks

Materials
- 10 cm lengths of 1.5 mm diameter glass tubing (GC 150F)

- 10 cm lengths of 2.0 mm diameter glass tubing (GC 200F)
- brass ferrules 10 mm long, 6 mm o.d. drilled to 4.5 mm i.d.
- Araldite epoxy resin adhesive (slow set)
- 20 × 20 cm square brass plate, 6 mm thick, with parallel V grooves 5 mm deep every 15 mm

Method

1. Select four lengths of the 1.5 mm tubing and one length of the 2 mm tubing.
2. Slide two brass ferrules onto the glass tubes.
3. Mix the Araldite resin and hardener and coat the glass tubes 15 mm from either end.
4. Slide the ferrules over the adhesive.
5. Place the blank into one of the grooves on the brass plate, press down, and rotate the ferrules in the groove.
6. When filled with blanks, put the plate into a 200 °C oven for 1 h, or alternatively, leave the blanks on the plate for 24 h.

The brass ferrules enable the glass to be gripped firmly by the chucks of the electrode puller and the electrodes to be held by the micromanipulators. It is sensible to choose an outer diameter for the brass ferrules that matches whatever arrangement exists on the micromanipulators for holding the electrodes.

2.1.1 Economies in the manufacture of blanks

It is possible to reduce the costs of the electrodes by buying glass tubing without capillary fibres included during manufacture. These are much cheaper, and capillary fibres may be inserted into the barrels just before the electrodes are pulled. The fibres may be handmade in a bunsen flame by pulling apart pieces of the 1.5 mm tubing. It is best to include two capillaries per barrel. These electrodes will have a higher failure rate than those described above.

It is also possible to omit the use of brass ferrules. The tubes may be held together with shrinking thermoplastic tubing intended for use in electrical circuits to bind harnesses of insulated wire together. One of the barrels should project beyond the others. This one is gripped by the puller and later is held by the manipulator during use. Designs of this sort work quite well but, in our experience, are not quite so dependable and show a distressing tendency to fall to bits at a critical moment. Remember that the drug solutions can be awesomely expensive; perversely, micropipettes seem to break after filling.

2.2 Pulling micropipettes

The puller used for these multibarrelled pipettes is arranged vertically. Two chucks of the type used in a small electric drill face each other, one above and one below a coil of nichrome wire. The electrode blank is placed into the upper chuck and lowered through the wire coil, so that it can be gripped by the lower chuck. The wire coil is carefully adjusted so that the glass is located centrally within the coil and the tubes will be heated evenly by the coil. Failure to do this leads to electrodes that are bent at the tip. The weight of the lower chuck, and whatever is hanging off it, is applying a continuous pull to the glass. It cannot be reduced with this type of puller, but may be increased by passage of a current through a solenoid. The AC currents heating the wire and passing through the solenoid are supplied by a double variable transformer. These currents are switched off and/or on by microswitches activated by movement of the lower chuck. Usually, the current to the heating coil is turned off when the glass is molten and elongating; the current to the solenoid is turned on at this point to pull the two chucks briskly apart. Pullers of this design are sold by Narishige (model PE-2M).

Using the puller to produce reliable multibarrelled micropipettes, which have good recording and iontophoresis characteristics, is an art which can only be obtained with practice. *Protocol 2* describes the sequence we use.

Protocol 2. Pulling micropipettes

Materials and equipment
- vertical electrode puller, e.g. Narishige PE-2M
- multibarrelled pipette blanks: *Protocol 1*
- horizontally mounted clamps to hold the electrodes when pulled
- stopwatch

Method
1. Place a blank into the upper chuck and clamp the chuck onto the brass ferrule.
2. Lower the upper chuck, raise the lower chuck, and clamp the ferrule.
3. Move the heating coil until it is midway between the two ferrules.
4. Check carefully that the glass is exactly concentric to the heating coil.
5. Turn on the heating current (1st phase) and leave for a timed period (60 sec).
6. Rotate the lower chuck, ferrule, and glass through 270 ° and allow them to fall for 5–10 mm.

7. Turn off the heat and allow the glass to cool for 1 min.
8. Lower the heating coil about 5 mm to ensure the thinnest part of the narrowed glass is heated.
9. Turn on the heating current (2nd phase) and allow the chuck to fall with a force of about 650 g.

The first heating phase is just sufficient to allow the glass to melt—the heating coil has an orange colour. The second phase is hotter, the coil is bright orange and the glass will melt after about 15 sec.

The preheating and then rotation of the glass causes the five tubes to weld together and, when pulled, one tip rather than five separate ones to form. The slight elongation of the glass at this time (the pre-pull) is where the art comes in. The principle is simple: the larger the volume of glass that becomes molten, the longer the pulled tip of the electrode will be (*Figure 1*). As explained later, this should be about 8–12 mm at most. If the pre-pull does not narrow the glass much, a lot of glass will contribute to a very long tip. Alternatively, if the pre-pull narrows the glass excessively, very short electrodes will result.

2.2.1 Profile of micropipettes

It must be remembered that a substantial proportion of the resistance of an electrode is longitudinal resistance and, the longer the shank, the more likely it will be to block when iontophoretic currents are passed. On the other hand, short stubby electrodes build up pressure in front of them as they are lowered into tissue. The short, low-resistance electrode may record cells and pass iontophoretic currents very well, but the cell will be 'lost' as the pressure ahead of the electrode dissipates with time and the tip drifts forward. Sharper electrode profiles will reduce this, but will have a higher resistance, therefore a compromise must be sought. It is usual to assume that high-resistance electrodes will record from a single cell rather than a population (see later). Although this is true, longitudinal resistance does not contribute to this ability to 'isolate' cells and it is better to think in terms of tip diameter. The smaller the tip diameter of the electrode, the more likely it is to record action potentials from one cell.

The glass at the tip of the electrode can become very fragile if heated too much. This may be because the wall of the tubes has become too thin, or because the heating and cooling processes have sintered the glass. This condition must be avoided at all costs, because the electrode will seem to be excellent electrically but will be very bad in use. Thus, the drugs will emerge along the length of the shank, not just at the tip. It can be seen that this is happening when breaking the electrode tips, as described below. The glass just crumbles away. The tips should be strong and resistant, i.e. rather difficult to break.

2.3 Preparing the electrodes for filling

2.3.1 Splaying the ends

The ends of the glass tubes are splayed apart by using the smallest gas flame you can arrange. We use the 'pilot' jet of a Bunsen burner. One by one the tubes are softened by the flame and with the tip of a needle they are splayed apart (*Figure 1*). This is very important because the filaments in the barrels will, by capillary action, draw the solutions in the barrels upwards (as well as downwards to fill the tip). The solution will evaporate at the top of the tubes and crystals will form at the end. If all the tubes and conducting crystals remain in contact with each other, then a current which is intended for one barrel may easily be passing down several other barrels. Multibarrelled electrode blanks are on sale which cannot be splayed apart and suffer badly from this problem.

2.3.2 Breaking back the tips

The tip of the electrode should be much less than 1 µm in diameter and will not be resolvable under a light microscope. Iontophoretic currents will not pass sufficiently, nor action potentials be recorded extracellularly from such tips. It is necessary to break the tips until they have a diameter of 5–7 µm. It is best to do this under visual control and not to rely on measurement of electrical resistance to give an indication of tip size. A bubble in the electrode is quite capable of causing an electrode with a large tip to have a high resistance.

A simple bench microscope with a fixed barrel and the usual manipulable table will suffice, but an unusually long working distance objective is necessary. We use one with a working distance of 10 mm and × 40 with × 10 eyepieces, and an intermediate magnification of 2.5. The overall magnification of 1000 resolves the tip of a 5 µm electrode, so that it may be measured to the nearest micron with an eyepiece graticule.

A 6 mm diameter glass rod is heated and pulled, and then heated again at the pulled part to form a small bead of 1–2 mm diameter. This is held in a manipulator and one face of the bead is brought into focus. The electrode is attached to the slide holder of the microscope and this is manipulated to bring the electrode tip into focus. The tip is then bumped against the glass bead, its diameter being monitored continuously, until 5–7 µm is achieved.

2.4 Drug solutions and electrode filling

If the main part of the glass tube is filled with a drug solution, the glass filament within the tube will rapidly convey the solution into the tip of the pulled part of the tube. An air bubble is left in the tube and this will float to the surface, especially if encouraged by tapping the electrode.

Drug solutions are injected from disposable 1 ml tuberculin syringes via

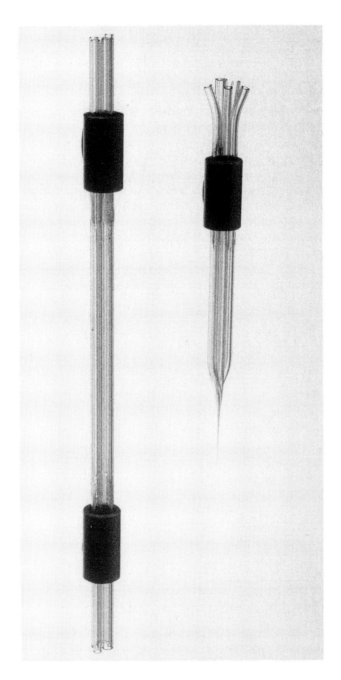

Figure 1. An electrode blank before pulling, and the completed electrode before filling with drugs. The manufacture of these is described in *Protocols 1* and *2*.

polythene tubing (initial o.d. 0.6 mm) which has been elongated in order to thin it. Care must be taken to insert only liquid into the barrel. It is very easy to insert bubbles as well. These will disturb the steady filling of the electrode tip, and remarkably bad electrodes can be produced by this small error.

2.4.1 Concentration of solutions

The drug solutions themselves can easily be the cause of electrodes that are noisy or fail to pass current. Obviously, if the solution inherently has a high electrical resistance, the electrode will have a poor performance even if it has been filled with skill and all bubbles excluded. To achieve good conductivity, it is often necessary to use a high concentration of the drug. Even with very expensive drugs it is sensible to use as high a concentration as possible. However, all drugs are different with regard to a wide variety of characteristics, and very few general rules can be made which apply to all. One such general rule, however, is that all drug solutions should be freshly prepared just before filling and the electrodes should be used as soon as possible after filling. Too many drugs are unstable and will degrade to make it worthwhile risking the use of electrodes more than 12 h old.

2.4.2 Volume of solutions

We find that little is gained by filtering drug solutions, providing care is taken to keep all containers and solutions dust free. The use of filters also increases the volume of the drug solution which has to be made and, with most solutions, this is impracticable due to expense. Each barrel of the electrodes we use holds 0.05 ml and, if four electrodes are prepared, 0.2 ml of solution is needed. We find that this is the smallest practicable volume which can be handled routinely without specialized equipment. This volume is only manageable if the drug is readily soluble in distilled water. Some drugs require extensive agitation, heating, titration, or pH adjustment before they will dissolve, and these manipulations are difficult with very small volumes. Heating the solution is in any case a dangerous process, which will significantly accelerate the degradation of most drugs. There are a few drugs which are sufficiently stable to be dissolved in this way.

2.4.3 Conductivity of solutions

The conductivity of a drug solution is extremely dependent upon the formation of charged ions. Thus, anything which increases the density of the charged ions will reduce the likelihood of the electrode blocking or otherwise failing in service. With most salts, the proportion of the ions carrying a charge varies with the pH. The value of pH which causes 50% of the ions to carry a charge is known as the pK. This value is available for most drugs, and the rule of thumb used by us is that the pH of the drug solution should be brought at least to the pK. It may be necessary, therefore, to adjust the pH of the solution. This can be done by adding the minimal volume of a strong acid or base, but it must be realized that this process will itself cause rather serious problems. When a voltage is placed across an ionized solution, current will flow, i.e. charged ions will migrate. If only the ions from the drug are present, then only these ions will carry the current. However, if the pH has been adjusted by addition of other ions, these also will migrate and will be carrying a proportion of the current. Fewer ions from the drug will emerge with any particular current than previously. This is the opposite of the effect hoped for from the pH adjustment. Adjustment of the pH is nevertheless necessary with some drugs because the electrode will otherwise conduct the current poorly. Do not assume, however, that the much freer passing of current is necessarily related to an improved emergence of drug ions from the electrode barrel.

Sometimes it is necessary to titrate an insoluble drug which is a base into solution. We have had to do this with an early formulation of the drug ketanserin which was available only as a base. Addition of a carefully calculated amount of tartaric acid created a solution of ketanserin tartrate without, of course, forming a highly acidic solution. Some drugs are zwitterionic; that is to say they may ionize in quite different ways depending upon the pH. Such drugs have two or more pK values, which are usually called the pK_a and the pK_b. It is necessary to decide carefully which form of ionization is to be used and to ensure that the pH is adjusted appropriately. Failure to do this may mean that the biologically active part of the drug molecule exists in the solution as both positively charged and negatively charged ions. This means that any polarity of current will eject the biological activity from the electrode. Adjustment of the pH will favour one type of ionization and inhibit the other. Glycine is zwitterionic, but at pH 3.5 the biologically active part of the molecule is mostly positively charged and can, therefore, be ejected by application of a positive charge and retained within the electrode by a negative charge.

The above brief survey of the factors to be considered when making up solutions for iontophoresis has assumed that the goal is to obtain maximal drug release. It should be noted that occasionally it is necessary to avoid this, when the potency of the drug is particularly high. With such drugs, it sometimes happens that either any application of any current gives supramaximal effects or the drug is effective even when not being applied, because of leakage from the electrode (see later). The solution to this problem is to dilute the drug, not with water but with competing ions. The addition of NaCl will significantly reduce the amount of drug emerging with any particular current.

3. The release of drugs from micropipettes

Drug ions will emerge from micropipette tips due to several different types of force. Hydrostatic efflux and diffusion will occur in the absence of any electrical charge, and both electro-osmosis and microiontophoresis occur when currents are passed.

3.1 Hydrostatic efflux and diffusion

Hydrostatic efflux will occur due to the pressure head imposed by the column of liquid contained in the barrel of the electrode. As this is only 1–1.5 cm of water and the orifice of the electrode is very small, hydrostatic efflux is usually considered to be negligible. Drugs that will not ionize in a charged form are sometimes ejected from the tip by pressure. Pressures in the region of 35–200 kN/m^2 are used to obtain responses from cells and measurable amounts of drug release.

Diffusion is a very different matter. It is very simple to demonstrate that this causes considerable efflux of drug. Although diffusion is not dependent upon the passage of an electrical current, it is considerably altered when a current is passed and it is possible to prevent diffusional efflux. This is done by passing a small current of about 25 nA which has the opposite polarity to that which affects the drug (the retaining current).

3.2 Electro-osmosis

Electro-osmosis was recognized a long time ago to be a major contributor to the movement of drug within electrodes, when a charge is applied to the top of the electrode. At the interface between the glass and the drug solution, charge separation exists, which tends to make the glass negative and the solution positive. Obviously, when a charge is applied to the top of the electrode, the fluid will move. As the nature of the glass, its history of heating and pullng, the concentration of the solution, and characteristics of both solvent and solute are all liable to contribute to the size of the charge in the solution and its mobility, predicting the contribution of electro-osmosis to the emergence of drug from microelectrodes is pointless. Too many assumptions have to be made to yield even an approximation. Experimental measures of the contribution of electro-osmosis to release from microelectrodes have been made, and it has been observed that it accounts for between 11 and 23% of the total release (1, 2). Electro-osmosis is, however, such an unpredictable phenomenon that it may account for much higher proportions of release with some drugs, solutions, or glass. This will be particularly so when the solution is concentrated because the whole solution moves with electro-osmosis, whereas with iontophoresis, cations and anions move in opposite directions and the bulk solution does not move.

3.3 Microiontophoresis

The last of the significant forces expelling drugs from glass microelectrodes is microiontophoresis itself. The greatest proportion of drug release is effected by this force. Theoretical calculation of the movement of ions is very easy and is described by Faraday's law (see above). This shows that there should be a linear relationship between the intensity of the ejecting current and the rate of iontophoretic release. Under practical conditions, however, this linearity may not hold, and release can be very variable. Bradshaw and colleagues, between 1973 and 1979, measured the release of drug under simulated experimental conditions and described a model of ion movements in micropipettes (Section 4). Attention to this model will avoid many of the serious mistakes which are the cause of the variable data in poorly conducted iontophoresis experiments.

4. Experimental measurement of drug release

Bradshaw *et al.* (3) filled microiontophoresis electrodes with ^{14}C-labelled noradrenaline of known specific activity and placed the tip of the electrode into saline. The radioactivity released into the saline was determined by scintillation counting. Currents could be passed through the electrode and their effect on drug release studied. Diffusional efflux was determined by studying release in the absence of any current.

4.1 Diffusional release

Diffusional efflux from five-barrelled pipettes of 4–6.5 mm diameter, containing 0.02 M noradrenaline bitartrate, was about 0.4 pmole/barrel/min, but was very variable. Greater diffusional release was seen from larger diameter electrode tips. Diffusional efflux from electrodes containing 0.2 M noradrenaline bitartrate was approximately ten times greater (4.9 pmole/barrel/min) than for the 0.02 M solutions. This efflux of considerable amounts of drug is prevented in biological experiments by the application of a retaining current. Retaining currents of 5 nA reduced diffusional efflux to about 25%, and 25 nA made diffusional efflux undetectable.

The release of noradrenaline during application of a range of ejecting currents was linearly related to the size of the ejecting current. As spontaneous release at 0 nA occurred, it was essential to subtract this spontaneous efflux from the total release before using Faraday's law to calculate the transport number. This spontaneous efflux, which can be as high as 30% of the total release, has another vitally important implication for biological experiments. Increasing the ejecting current will not *pro rata* increase the rate of release from the electrode, an assumption that has been the basis of many attempts to 'quantify' microiontophoresis. This assumption cannot, therefore, be valid.

4.2 pH and the effects of competing ions in the solutions

Bradshaw and Szabadi (4) also investigated the effects of diluting the ions of the drug with other charged ions. The result is that the iontophoretic current will be carried not only by the cations and anions of the drug solute, but also by others. The precise effect of this will depend upon the valency, concentration, mobility, etc. of the diluting ions. Bradshaw and Szabadi chose a very practical example. Noradrenaline bitartrate (0.2 M) in distilled water has a pH of 3.5. This pH has been of concern to some authors, who felt that artifacts due to pH would be less likely if a less acid medium was used. The pH was adjusted to 5.0 by the addition of NaOH. The concentration of Na^+ in the adjusted solution was 0.1 M. There was no difference in the spontaneous efflux of radiolabelled noradrenaline from the two solutions. Retaining and ejecting currents were less effective in the solution at pH 5.0. The apparent transport number (0.2 M noradrenaline) at pH 3.5 was 0.29 and at pH 5.0 was 0.11. Clearly, addition of competing ions, especially small, highly ionized, and mobile ions like Na^+, causes loss of control of the drug ion during iontophoresis experiments. Again, the magnitude of the effect is so great that the point should be considered very seriously by anyone using the technique.

4.3 Retaining currents

Bradshaw and colleagues (3–5) showed also that the retaining current, its duration, and its magnitude, grossly affected the subsequent release of drug during a period of ejection. The essence of the phenomena is revealed by *Figures 2* and *3*. These show that the release of drug (*Figure 2*) and the response of cortical neurones (*Figure 3*) both depend upon the retaining current. A long period of retaining current application or a large retaining current profoundly reduces both the drug release and the neuronal response.

In *Figure 2*, drugs were ejected for periods of 8 min, which is far longer than the periods normally used in biological experiments. This was done because the specific activity of the available noradrenaline was rather low. It is to be expected that shorter sampling times would reveal the effects in a very much more dramatic manner.

The histogram that forms the upper part of the figure shows the noradrenaline released from the electrode during 8 min sampling periods. The ejecting and retaining currents are shown in the lower histogram. At first, the noradrenaline is simply allowed to diffuse into the collection vials. This diffusion was almost completely prevented by application of a 25 nA retaining current. After 64 min of constant application of this retaining current, alternating 8 min periods of ejecting and retaining currents were studied. Ejecting currents were invariably an 8 min period of 50 nA. Only the retaining currents were varied. This variation in the retaining currents was entirely responsible for the very considerable variation in the release of noradrenaline from the electrode. Following the long period of retention (which is equivalent to searching for a cell in biological experiments), the first release period ejected rather little drug. Subsequent ejections released increasing amounts of the drug. At the point labelled 'C', the retaining

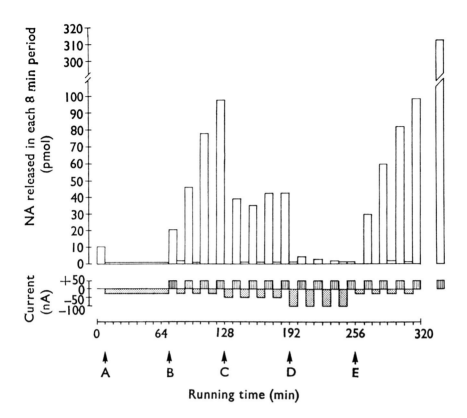

Figure 2. The effect of changing the duration and intensity of retaining currents upon the electrophoretic release of noradrenaline (NA) from four barrels of a micropipette. The lower panel shows the electrophoretic current applied to each of the four barrels. Downward deflections represent retaining currents; upward deflections represent ejecting currents. Note that the duration and amplitude of the ejecting currents do not change throughout the experiment. Changes in the parameters of the retaining current are indicated by the capital letters under the time base. The upper panel shows the release of NA during each 8 min sample period. At first, 10 pmol of NA emerged during the collection period when no ejecting or retaining current was applied. During the prolonged application of 25 nA of retaining current (A–B), the release was nearly, although not quite, abolished. Regularly alternating applications of ejecting and retaining currents (B–C) gave rise to progressively greater release of drug during the ejection periods—drug release 'wound up'. Increases in the intensity of the retaining current reduced the amount of NA released by the ejecting pulse (C–D, D–E). Restoration of the original retaining current at (E) was followed by progressively increasing outputs. To the right is shown the release caused by a standard ejection pulse which is not preceded by a retaining current. These data demonstrate that drug release is dependent upon the recent history of the electrode and that timers must be used to ensure that intervals between drug applications do not vary. (Reproduced from reference 3, with permission.)

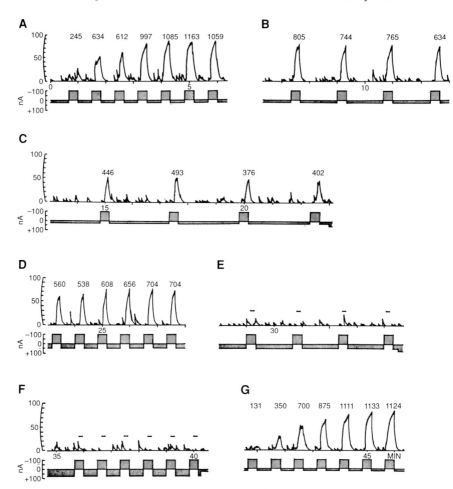

Figure 3. The effect of changing the parameters (intensity and duration) of the retaining pulse upon the responses of a rat cortical neurone *in vivo* to microelectrophoretic applications of glutamate. The intensity and duration of the ejecting pulses were not varied. Upper traces: Continuous recording of the firing rate of a single cortical neurone. Figures above the traces show the total number of spikes generated in response to each application of glutamate. Lower traces: Electrophoretic currents. Before the start of this study, the 25 nA retaining current had been applied for 3 min. The first ejection caused little response, but subsequently the response built up to a stable plateau (A). In (B) the effect of increasing the period of drug retention between the standard ejections is shown; responses are smaller. Further increases in the retaining period (C) causes further reduction of the response. In (D) the period of retention is the same as in A but the size of the retaining current is 50 nA; responses are smaller than in A. (E) and (F) reveal that further increases in duration or amplitude of the retaining current practically abolish responses from this cell. It is shown in (G), however, that the original response can be restored by returning to the original response-retaining currents. It is clear that variation in response amplitudes (and drug release—see *Figure 2*) are to be expected, if drugs are ejected in a haphazard way with non-standard intervals. (Reproduced from reference 5, with permission.)

current intensity was increased to 50 nA and this more than halved the drug emerging during the ejection periods. At 'D' the retaining current was increased to 100 nA and this practically abolished any release of drug from the electrode. Returning the retaining current to 25 nA, again allowed the drug release to build up progressively. Finally, after an extended period with no retaining current applied, the standard ejection released a very large amount of drug.

In practical terms, this means that long periods of retaining current may abolish completely the release of drug during the first period of intended application. Subsequent periods of application may be separated by only short periods of retention and, therefore, progressively more drug may be released until stability of release is achieved. This may take a long time (see *Figure 3*). The response of central neurones to an excitant drug will apparently 'wind up' due to this phenomenon.

These phenomena are not only seen in specially designed release experiments. They are also seen in biological experiments where failure to understand the changes in drug release that are occurring have led to mistaken conclusions.

An example of changes in neuronal response is shown in *Figure 3*. The records are continuous and were obtained from a single neocortical neurone in the fluothane-anaesthetized rat. The upper trace in each panel shows the neuronal firing rate, and the lower trace shows the magnitude and direction of the iontophoretic current passed. The iontophoresis electrode was filled with sodium glutamate (pH 7.5; 0.2 M), which is, of course, ejected from the electrode by a negative charge. At the start of the experiment, a retaining current of +25 nA had been applied for some time. An ejecting current of −100 nA was applied for 20 sec at intervals throughout the study. This intensity and period of drug application was not varied at any stage. The first ejection evoked little activity but, with repeated application at 30 sec intervals, the response apparently 'wound up' until a stable response was achieved. This stability is revealed by the numbers above the responses which show the total number of spikes evoked by each application of glutamate. Increasing the interval between ejection periods without altering the retention current reduced the size of responses markedly. In panel D, the original periods of ejection and retention were reinstated but the size of the retaining current was doubled. The responses were much smaller than in panel A, however. When the duration of the 50 nA retaining current periods was increased, responses to 20 sec 100 nA ejections of glutamate were absent. Reinstatement of the original retaining current intensity and duration caused exact reinstatement of the original response amplitude, although time was taken for these responses to recover. These data (*Figures 2* and *3*) show that microiontophoresis experiments must be conducted in an extremely systematic manner. Varying the interval between agonist applications causes predictable and very large changes in response amplitude. If an unusually long gap between agonist applications occurs when an antagonist is applied, the response to the agonist will be much smaller, even when the antagonist is ineffective.

4.4 Bradshaw's model of ion movements in a pipette

Bradshaw and his colleagues (4) developed a model which assists with the conceptualization of the changes in dilution at the tip of the electrode during iontophoresis experiments. Several elements of the model had been less clearly introduced previously (6). When an electrode is broken back to approximately 5 μm and immersed in water or brain tissue, diffusion of anions and cations out of the electrode barrel occurs. This eventually causes a very low concentration of drug to exist in the tip. A progressively higher concentration exists higher up the shank of the electrode. Application of a retaining current pulls the biologically active ion back up the shank of the electrode and away from the tip, which, in time, may come to contain none of these ions at all. It is only at this time that the retaining current becomes completely effective. Continued application of the retaining current will pull the biologically active ions increasingly further up the shank. When an ejecting current is applied, the ions will start to move down the shank towards the tip. They will take time to do so, however, and for a period no biologically active ions will emerge. When they start to do so, it will be very slowly at first, progressively accelerating to maximum when the ion concentration in the bulk of the solution arrives at the tip. Once this has occurred, diffusional release as well as iontophoretic release will be maximal and will plateau. All these processes will be relatively rapid with small and highly mobile ions, but will take much longer with larger ions which will have small transport numbers. The practical relevance of the above sequence is illustrated by *Figure 3*. Glutamate is one of the most mobile of the drug molecules and yet, as the figure shows, with quite normal periods of retention and ejection, with currents well within the ranges used in experimentation, the response will vary from large to absent. Less mobile ions like noradrenaline or 5-hydroxytryptamine will be more affected by the retaining current and require more time for equilibration. Very large ions like some of the peptides may require very long periods of retention before diffusional efflux is prevented and, subsequently, very long periods of ejection before any drug emerges. It is an arguable point that the larger the biological ion, the less useful microiontophoresis has been, or will be, in revealing its effects.

5. Practical experimental design

The Bradshaw model which has been outlined above is presented in much more detail in the original articles, which also demonstrate fully that the concepts have been validated experimentally. Consideration of these principles and other, more widely known, concepts in pharmacology and single cell recording enable the creation of effective experimental designs when using microiontophoresis. The most important limitations in addition to those outlined above are as follows.

The cell–electrode distance is not known in iontophoresis experiments. No clue as to distance from the cell can be obtained from action potential size and, in any case, receptors for the iontophoretically applied drug may exist on dendrites; these may be very close to or distant from the electrode tip. This unknown distance, however, is crucial to drug concentration. The greatest concentration of drug ions during iontophoretic ejection exists at the tip of the electrode. The drug will diffuse outward from this point. Assuming (naïvely!) a homogenous medium, the diffusion will create spherical shells of decreasing concentration with increasing distance from the tip. A sharp exponential decrease in concentration may be expected. Curtis and Watkins (7) made theoretical calculations and proposed that an approximately eightfold difference in concentration may be expected at distances of 20 and 40 μm from the tip. This difference becomes less as ejection of the drug continues, but is still threefold after 100 sec of iontophoretic ejection, and does not decrease significantly further when time = infinity. It may confidently be concluded that the concentration of drug at receptors is unknown during iontophoresis experiments. It is wrong, however, to conclude as many authors have that no form of quantification is possible with iontophoresis (see below).

Release experiments conducted with radiolabelled drugs ejected iontophoretically from micropipettes have been done by very many different laboratories. They all show that a surprisingly high degree of variability occurs between electrodes, even when the electrodes are matched for size and electrical resistance. It really is a very difficult matter to contribute accurate data on dose–response relationships with the technique of microiontophoresis, but it is just possible with great care. In reality, the problems with the microiontophoresis technique are surprisingly similar to problems with other, more classical, pharmacological techniques. Most pharmacologists are comfortable with *in vitro* (organ bath) techniques where precisely known concentrations of drugs are added to the bath. However, our knowledge of the concentration of drug at the receptor in these experiments may be as flawed as with the iontophoresis technique. The receptors may be on the surface or deep in the tissue. The speed of penetration of drug into the tissue is difficult to determine and is very different for different drugs and different tissues. The concentration of drug at different depths within the tissue will be different, therefore, and will bear an unknown relationship to the concentration of drug in the bath. Plateau responses following long exposures to the drug are often spoiled by tachyphylaxis at the receptors. Of course, in spite of these very real difficulties, organ bath techniques have made enormous contributions to pharmacology and so has the technique of microiontophoresis. The rapidity of drug ejection from microelectrodes reduces the likelihood of tachyphylaxis obscuring a receptor-mediated response, and the direct application into the extracellular environment avoids the slow and uncertain degree of drug penetration (especially through the blood–brain barrier) that is encountered with other methods.

5.1 A recommended method—quantification of drug release

A semi-quantitative method of comparing agonist potencies was used by Davies and colleagues (8). The method is based upon the knowledge that the variance of drug release from barrel to barrel of the same micropipette is much smaller than the variance of drug release between different micropipettes. It is necessary to determine the transport number of each of the biologically active agonists that are to be studied. This is done as described in *Protocols 3* or *4* (9).

Protocol 3. Determination of the transport number of a radiolabelled drug with a high specific activity

Materials
- microelectrodes
- radiolabelled drug solution in distilled water[a]
- scintillation vials (10 ml) containing 0.5 ml normal saline

Method

1. Fill four barrels of each of five electrodes with the radiolabelled drug solution.
2. Take the first electrode, connect to the iontophoresis panel and place a scintillation vial below it. Earth the saline in the vial.
3. Wash the electrode tip with distilled water, discard the washings, and lower the tip into the saline in the scintillation vial.
4. After 5 min, raise the electrode, replace the vial, wash the electrode, and lower into the new vial. Do this as quickly as possible.
5. Repeat step 4 at least three times to determine release at 0 nA iontophoretic current.
6. Repeat step 4 at least a further three times, but apply a current of 25 nA to each barrel during the 5 min collection period.
7. Repeat step 4 at least a further three times while applying currents of 50 nA and 100 nA to each barrel.
8. Repeat steps 2–7 with each of the other four electrodes.

9. Place the scintillation vials into a counter and determine the drug release in picomole/min/barrel.
10. Graph the mean drug release for each current and calculate the transport number of the biologically active ion as follows.
11. Fit a straight line to the graph and determine the release at 0 nA (diffusional release).
12. Subtract this from the release at 100 nA (iontophoretic release).
13. Calculate the transport number (n) from the equation:

$$n = \frac{RizF}{i}$$

where z = valency of the biologically active ion, F = Faraday's constant = 96 500 coulombs, Ri = iontophoretic release in mole/sec, i = iontophoretic current in amps.

[a] It is not possible to specify the exact specific activities, times, and other quantities as these will be quite different for different drugs. Obviously, however, these quantities must be sufficient to raise the counts obtained to well above background levels for even the smallest releases.

Many compounds, especially the newer and often most useful ones, are not available as suitably radiolabelled molecules unless very high prices are paid. Bradshaw and colleagues (9) have developed a competition method that makes such expense unnecessary. A solution is made of the radiolabelled agonist as described previously, and then a 50:50 mixture is made of the radiolabelled agonist and an identical solution of the new non-labelled agonist. This mixture is placed into two barrels of a multibarrelled pipette and the single solution of radiolabelled agonist is placed into two other barrels of the same pipette (*Protocol 4*).

Protocol 4. Determination of the transport number of a drug not available in a suitable radiolabelled form

Materials
- all the materials detailed in *Protocol 3*, including any drug which is suitably radiolabelled
- the experimental drug[a]

Protocol 4. Continued

Method

1. Make 1.0 ml volumes of a solution of each of the radiolabelled and the experimental drugs. The concentrations should be identical and the same as those used in biological experiments.
2. Fill the electrodes with solutions as follows:
   - two barrels with the radiolabelled solution
   - two barrels with a 50:50 mixture of the radiolabelled and experimental drug
3. Proceed as in *Protocol 3* using steps 2–5.
4. Repeat step 3 (steps 2–5, *Protocol 3*) but apply an ejecting current of 25 nA to each of the two barrels containing the radiolabelled drug. Repeat three times.
5. As step 4, but apply 25 nA ejecting currents to each of the two barrels containing the mixture of radiolabelled and experimental drugs.
6. Repeat steps 4 and 5 but apply 50 nA ejecting currents three times and then 100 nA ejecting currents three times.
7. Graph the drug release in picomoles of the radiolabelled drug/min/barrel.
8. Using the data from the barrels containing only the radiolabelled drug, calculate the transport number of the radiolabelled drug (see *Protocol 3*).
9. Using the data from the barrels containing the mixture of the two drugs, calculate the transport number of the radiolabelled drug.
10. Note that if the transport number of the radiolabelled drug is halved by the 50:50 dilution with the experimental drug, then both drugs have the same transport number.
11. Note that if the transport number of the radiolabelled drug is more than halved, the experimental drug is more mobile than the radiolabelled drug. The relationship between iontophoretic release and transport number will be linear and the transport number of the experimental drug can be calculated.

[a] The procedure assumes that both drugs are acids or bases, or are salts of the same acid. It must be remembered that the counter-ion carries a proportion of the charge and, if the counter-ions have different mobilities, this may account for the perceived differences in release of the drugs. It is possible to check if the mobilities of different counter-ions are different, but it is better to choose a radiolabelled drug which has the same counter-ion.

5.2 Determination of comparative potency in biological experiments

Let us assume that the observation concerning the mobilities of the two agonists indicates the simplest case, i.e. that they have identical mobilities. Under these circumstances, different barrels containing the same concentration of either agonist will show a very similar average release. Although the concentration of either agonist at the receptors during iontophoresis experiments will not be known, it can be confidently predicted that, whatever that concentration is following application of a particular ejection current and time, it will be similar for both agonists. If the response of a cell to one agonist is much larger than the response of the same cell to the other agonist, which was identically ejected from the same electrode, then it must be concluded that the cell is more sensitive to one of the agonists. It is likely that one agonist is more potent than the other. This may be because the receptors have a greater affinity or that one drug has a greater efficacy. However, do not forget that just as with any other pharmacological technique, an apparently greater potency may be due to differences in the degradation of the two drugs or to other factors influencing bio-availability.

When the biological observations have been replicated sufficiently, it will be possible to calculate meaningfully the average difference in potency of the two drugs. The variability of this difference in potency, as well as the variability in the release of drug from different barrels of the electrode, may be calculated and a level of significance may be attached to the observation. As receptor types are to some extent defined by potency ratios, this method enables the identification of receptor subtypes on central neurones. For the potency ratio to be meaningful, it is, of course, necessary to show from other experiments that the two agonists elicit responses by acting upon the same receptors. This usually requires antagonism of both responses with highly selective antagonists.

5.3 The control of extracellular current artefacts

There are many other aspects of practical design of microiontophoresis experiments which are easily understood, and have been employed by many authors for decades. Not all publications reveal an awareness of them, however, and seriously misleading data are produced quite regularly in the journals. It is necessary to remember that microiontophoretic applications involve two significant changes in the environment of neurones. Electrical currents are caused to flow and drug molecules appear. Obviously, excitable cells are very sensitive to current flow, and controls are needed to establish if a change in the excitability of the cell is in response to the current or to the drug. Central neurones often respond vigorously to the application of iontophoretic currents from barrels filled with sodium chloride. It is necessary to reduce these changes in current flow. There are two methods of doing this, as follows.

One method takes all the currents flowing through each barrel of the micropipette and ensures that they add up to zero. In other words, if the currents flowing through all the drug barrels add up to -45 nA, a current of $+45$ nA is applied to a sodium chloride barrel (the 'balancing' barrel). In our experience, this system is less than ideal and artifactual responses from cells are quite often seen. This may be because, when a drug is ejected, there is a sudden change from a small retaining current to a large ejecting current of the opposite polarity. This will be accompanied, of course, by similarly large changes in current flow from the compensating sodium chloride barrel. These currents must flow through the extracellular fluids, as only here are the barrels resistively coupled. The cells are, therefore, subjected to a sudden increase in extracellular current flows, even though at all times the net flow adds to zero.

The 'equal current' method of current balancing is our preferred method (10). In this system, the net current flow is not zero, but does not change when a drug is applied. If it is intended to apply a drug ion with a current of $+100$ nA, then during the periods preceding and following the drug application, a current of $+100$ nA is ejected from the sodium chloride barrel. During the drug application, the positive current from the balancing barrel is turned off and a positive current applied to the drug barrel. This technique is a little more clumsy and presents difficulties when agonists need to be applied alternately with different ejection currents. Confirmation of its (mostly) effective control of current artefacts is obtained when the system is used in association with intracellular recording from a brain slice. The 'net flow to zero' system causes considerable DC shifts in membrane potential to be recorded from a cell, whereas the 'equal current' method usually prevents such artifacts. Whichever system is used, it is essential to remember that, even with current balancing, occasional responses will be encountered which are due to current. Often these are indicated by the suddenness of the change in firing frequency or by a sudden change in spike height. However, very potent drugs which are highly mobile (and therefore rapidly ejected by iontophoresis) may also have sudden effects.

It is essential to apply unbalanced changes in Na^+ or Cl^- current from the balancing barrel to detect the sensitivity of the cell to deliberately applied change in current.

5.4 Changes in amplitude of action potentials

Extracellularly recorded spike amplitude must be expected to change in parallel with drug-induced changes in cell excitability. A decrease in spike amplitude often occurs with depolarizing drugs, which reduce membrane resistance, and the opposite often occurs when firing frequency is reduced.

Spike amplitude must be monitored with great care to ensure that apparent reduction in firing frequency is not caused by an actual reduction in spike amplitude. During extracellular recording, action potentials are usually passed through a voltage gate to a ratemeter. Action potentials will not, therefore, drive the ratemeter if their voltage drops below the gate. It is essential to be able to determine that the spikes are triggering the gate. The best way is to take an output signal from the action potential processing unit and display it on a second channel of the oscilloscope. Each counted spike should have above it the gated output. It is easy to arrange that the action potential triggers a storage oscilloscope. There is then usually plenty of time to see that the two signals always coincide (see *Figure 4*) and nothing but action potentials triggers the ratemeter.

5.5 Signal-to-noise ratio

The above requirement explains why experimental design requires a good signal-to-noise ratio, i.e. why a large action potential should be well separated from background noise. Modern amplifiers make this relatively easy to achieve, and this account does not detail different microelectrode recording methods. It is worth explaining why, when using multibarrelled microelectrodes, a good signal-to-noise ratio is much more difficult to obtain than usual. The reason for this is twofold:

(a) The drug barrels are closely capacitance-coupled to the recording barrel and, through the tissue, resistively coupled to the recording barrel. Any noise picked up by the iontophoresis wires, or injected by the iontophoresis control unit, will be readily recorded. This tends to make the noise level significantly higher than when single-barrel microelectrodes are used.

(b) The tip size of a five-barrelled (or worse, a seven-barrelled) microelectrode is very much larger than a single-barrel recording electrode. It is a well-known principle that, within limits, the finer the tip of an extracellular electrode, the more focal the recording; i.e., a nearby cell will produce a larger action potential from a fine-tipped electrode. Furthermore, a fine-tipped electrode will be able to get closer to a neurone without damaging it. On both counts the multibarrelled pipette may be expected to record relatively small action potentials. Signal-to-noise ratios are, therefore, rather poor, and gains of about 100 µV/cm are required, together with variable in-line filters which can be adjusted to optimize the spike amplitude.

5.6 The electronic timing of drug application

The theoretical model of release of drug ions from micropipettes, described by Bradshaw and colleagues (4), explains why the design of experiments requires several periods of drug ejection before a stable response can be

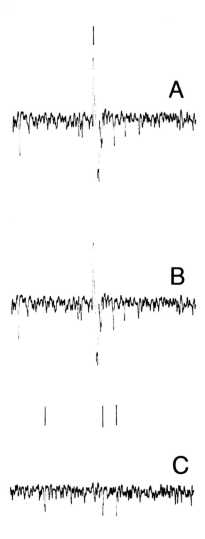

Figure 4. The method of displaying action potentials on an oscilloscope to ensure that all action potentials and only action potentials are being counted. (A) The action potential is used to trigger the time base of a digital storage oscilloscope. Above the action potential is a feedback pulse from the spike processor which shows that this potential has been counted. (B) The same action potential is displayed, but the pulse height discriminator has been altered; it can clearly be seen that the spike has not been counted by the equipment. (C) The equipment has again been incorrectly adjusted and small spikes are being counted. This method of using a feedback pulse above the counted signal reduces the likelihood of pharmacological studies of circuit noise.

expected. The retaining current depletes the electrode tip of drug ions and the first few ejection periods have to push these ions into the tip before they start to emerge into the environment of the cell. If an insufficient number of ejection periods are applied, it may be concluded that the drug is ineffective. Always allow for many repeats before identifying the action of any drug. It is vital to control these periods with electronic timers, as quite small variations in the interval between ejection periods will profoundly alter the amount of drug released from the electrode and destroy the apparent stability of the response. It is tempting to try to prevent this trouble by omitting to apply retaining current. This is always a mistake and can be a very serious one. The drug continues to diffuse from the electrode tip after the ejection period and tachyphylaxis will easily cause the desensitization of labile receptors, thus negating the major benefit of the iontophoresis technique. We use retaining currents of 15 nA for most drugs, but smaller currents down to about 5 nA may be used for highly mobile ions like γ-aminobutyric acid or glutamate. These low currents do not stop drug diffusion from the electrode completely, and checks must be made for the possibility of desensitization. Even with large retaining currents, it is still possible to desensitize some receptors by too frequent an application of the drug. The remedy is obvious and desensitization can easily be detected. When more time is left between applications, larger intervals may reduce drug release from the electrode. Thus, if responses nevertheless become larger, it is not due to a change in drug release but is due to a lessening of desensitization.

References

1. Krnjevic, K., Laverty, R., and Sharman, D. F. (1963). *Br. J. Pharmacol.*, **20**, 491.
2. Bevan, P., Bradshaw, C. M., Pun, R. Y. K., Slater, N. T., and Szabadi, E. (1979). *Br. J. Pharmacol.*, **67**, 478.
3. Bradshaw, C. M., Roberts, M. H. T., and Szabadi, E. (1973). *Br. J. Pharmacol.*, **49**, 667.
4. Bradshaw, C. M. and Szabadi, E. (1974). *Neuropharmacology*, **13**, 407.
5. Bradshaw, C. M., Szabadi, E., and Roberts, M. H. T. (1973). *J. Pharm. Pharmacol.*, **25**, 513.
6. Curtis, D. R. (1964). *Microelectrophoretics in Physical Techniques in Biological Research*, Vol. 5A (ed. W. L. Nastuk), pp. 144–190. Academic Press, New York.
7. Curtis, D. R. and Watkins, J. C. (1960). *J. Neurochem.*, **6**, 117.
8. Davies, M., Wilkinson, L. S., and Roberts, M. H. T. (1988). *Br. J. Pharmacol.*, **94**, 483.
9. Bradshaw, C. M., Pun, R. Y. K., Slater, N. T., and Szabadi, E. (1981). *J. Pharmacol. Methods*, **5**, 67.
10. Roberts, M. H. T. and Straughan, D. W. (1967). *J. Physiol.*, **193**, 269.

12

Electrophysiological techniques for studying the quantitative pharmacology of the isolated spinal cord

D. I. WALLIS

1. Introduction

The spinal cord from the neonate mammal provides a useful *in vitro* preparation for investigating the physiology and pharmacology of motoneurones and the pathways that converge onto them. This chapter gives examples of isolated spinal cord preparations from the neonate rat and illustrates the types of experiment that can be conducted. Because the spinal column is cartilaginous in the neonate, the cord can be removed rapidly without damage to either cord or roots. The survival of central nervous system (CNS) tissue is enhanced *in vitro* if:

- the tissue is superfused at an adequate flow rate with an oxygenated medium
- physical size is small, since distance over which oxygen has to diffuse from the superfusion medium is correspondingly small

and, optionally,

- if oxygen usage is reduced by lowering the temperature

We find that isolated cord preparations survive best at temperatures between 20 and 25 °C (see also Chapter 9).

Cord preparations can be hemisected to improve oxygenation of deeper tissues. However, by using animals of 1–2 days old, when the cord is very small, it is possible to set up more complex preparations consisting of whole cord without hemisection. These can be used to study bilateral reflex responses and preparations of cord with tail attached.

2. Extracellular recording of population responses

For quantitative pharmacological studies, it may often be preferable to record responses from populations of neurones, principally because of the difficulties in intracellular studies of holding impaled cells for sufficiently long to complete investigations with antagonists. Further, the fluctuations in excitability characteristic of individual cells in CNS tissue, which are due to neuronal activity, are to some extent averaged by sampling population responses. This leads to increased reproducibility of responses. This section deals with some extracellular recording techniques applied to the cord *in vitro*.

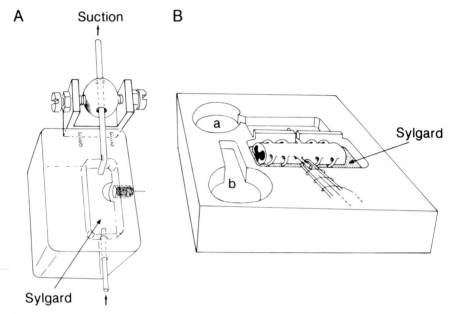

Figure 1. Perspex superfusion baths used for neonate rat hemisected spinal cord. Bath A (viewed from above) is used in conjunction with suction electrodes placed on dorsal and ventral roots. The block has a milled oblong chamber partially filled with Sylgard resin. Superfusion medium is introduced via the cannula (bottom) set at a level just above the surface of the Sylgard and removed by the adjustable steel tube (top) attached to a suction line. A hole bored on the left-hand side of the chamber carries a silver chloride pellet which serves as the reference electrode. Bath B is used to record changes in DC potential arising between the distal end of a ventral root and the spinal cord. Changes in potential arise in neurones (i.e. motoneurones) whose axons traverse the ventral root. The bath was milled from a Perspex block. The oblong chamber partially filled with Sylgard resin contains the hemisected cord. A dorsal root is drawn into a glass suction electrode. The corresponding ventral root is drawn through a narrow slot in a Perspex partition into an adjacent saline-filled chamber. The slot is sealed with a liquid paraffin and Vaseline mixture. The saline in the adjacent chamber is contiguous with saline in chamber 'a' into which dips a calomel half-cell. The chamber containing the cord is linked via a saline wick to chamber 'b' into which dips a second calomel half-cell. Thus, one calomel electrode makes electrical contact with the cord and central end of the ventral root, while the other leads from the distal end of the ventral root. The Vaseline-filled slot reduces short-circuit current flowing between these points. The cord is continuously superfused with oxygenated Krebs' solution.

2.1 Baths

Baths of simple design are perfectly adequate for neonate cord preparations (*Figure 1*). Bath A, or a slight modification of it, has been used for the experiments described in *Protocols 2, 4, 5,* and *6*; Bath B has been used for those described in *Protocol 3*.

2.2 Preparation of cord

Protocol 1 describes how to make a hemisected spinal cord preparation. In the UK, the procedure using neonate rats requires Home Office approval by licence.

Protocol 1. Preparation of hemisected lumbar spinal cord

Materials

- modified Krebs' solution (*Table 1*)
- Williamson–Noble iris scissors (John Weiss & Son)
- small Vannas iridectomy scissors (John Weiss & Son)

Method

1. Select a male or female 3–8-day-old rat. The size may vary considerably between litters from different parents, etc. Experience helps in selecting a rat of suitable size.
2. Deeply anaesthetize the rat with a suitable volatile anaesthetic, such as ether.
3. After decapitation, quickly remove the entire vertebral column and place in a shallow dissecting dish of oxygenated Krebs' solution or artificial cerebrospinal fluid (aCSF). Survival of neurones may be enhanced if the Krebs' solution is chilled.
4. The cord is encased by soft cartilage at this age. Transect the cord between two vertebrae in the cervical region.
5. Beginning in the midline on the ventral surface, carefully free the cord from the vertebral column using fine scissors.
6. Cut the exposed spinal roots as they enter the column to allow the cord to be floated free using iridectomy scissors.
7. Pin the cord into the Sylgard base of the dissecting dish using one pin at either end.
8. Remove the dura using iridectomy scissors and hemisect the cord sagitally using a Swann–Morton size 15 or 22a scalpel.

Protocol 1. *Continued*

9. Transfer the hemisected preparation to the recording chamber. Secure it at the cranial end with a pin if required. The Sylgard-filled trough (Bath A) has a volume of about 1 ml.
10. Immediately begin superfusion of the cord with oxygenated modified Krebs' solution (or aCSF) at 23–25 °C with a reasonably fast rate of flow. 2–3 ml/min seems to be adequate (1).
11. Alternatively, allow the preparation to recover from surgical trauma for an hour in oxygenated Krebs solution at room temperature (20–22 °C), before transfer to the recording chamber.

Table 1. Composition of modified Krebs' solution

Substance	Concentration (mM)	Take for 1 litre of Krebs'
NaCl	118.2	6.9 g
KCl	3	0.22 g
$CaCl_2$	1.2	1.2 ml of 1 M Analar solution
$MgSO_4.H_2O$	0.6	0.15 g
KH_2PO_4	1.2	0.16 g
Glucose	11.1	2 g

Add distilled water to give a volume of about 900 ml and equilibrate with 5% CO_2/95% O_2 for 10 min and agitate before adding bicarbonate.

$NaHCO_3$	25	2.1 g

Total molarity 160.3 mM, pH 7.44, when equilibrated with 5% CO_2/95% O_2.

2.3 Extracellular methods of recording

The arrangement of recording and stimulating electrodes for the hemisected cord is described in *Protocol 2*.

Protocol 2. Recording and stimulating

Materials

- glass capillary tubing: Microcap
- micromanipulators: MM3 or Prior, (Micro Instruments)
- microelectrode holder: MPH-1A (Clark Electromedical)

Method

1. Prepare glass suction electrodes from Microcap tubing or other suitable glass capillary tubing.

2. Anneal the ends of Microcap tubes to provide a range of orifice sizes to be matched with the spinal roots used. An adequate fit of a root in the suction electrode will cause slight dimpling of the nerve surface. Observe the fit, which should be sufficiently tight for the root not to slip out, but not so tight as to cause damage.
3. Make a recording electrode by introducing a chloride-coated silver wire into the prepared Microcap tube and cement in place using Rapid Araldite.
4. Make a stimulating electrode by introducing a silver wire into the prepared Microcap tube and seal in place, after winding an insulated silver or copper wire (with tip exposed) around the outside of the Microcap tube.
5. Normally, segment L_4 or L_5 ventral roots are suitable for recording. Choose a recording electrode of suitable size for the root you have selected. Mount the electrode in a micromanipulator (e.g. MM3 or Prior) and attach by polythene tubing to a 1 ml syringe to provide suction. Alternatively, hold the Microcap tube electrode in an MPH-1A microelectrode holder; attach a suction line and syringe to the side arm.
6. Apply a stimulating electrode to the dorsal root from the same segment. Select a stimulating electrode of suitable size. Mount the electrode in a second, similar micromanipulator.
7. Consult *Figure 2* for the apparatus required to superfuse the cord, to stimulate a dorsal root, and to record from a ventral root, etc.
8. Use stimulus pulses of 0.1 msec duration, 10–20 V to evoke a maximum monosynaptic reflex (MSR) and maximum polysynaptic reflex (PSR).
9. Record the complete response transient from the digital store of an oscilloscope and feed to an x–y plotter or Gould Color writer. Record changes in DC potential on a chart recorder running continuously at a slow speed.

2.3.1 Reflex responses

The reflex responses recorded arise in motoneurones whose axons run in the ventral root selected for study. Changes in stimulus intensity, and neurotransmitters such as 5-hydroxytryptamine (5-HT), which modify transmission, alter the amplitude of MSR and PSR. Thus, the automation of the measurement of the amplitudes and the continuous display of analogue MSR and PSR amplitude are very useful if quantitative pharmacological experiments are to be attempted. A peak height detector (2) can be used, modified to provide a double window or gate to sample the peak voltage occurring during the time window. The peak height detector is triggered by the stimulator/timer, and incorporates both sample and hold circuits and adjustable delays.

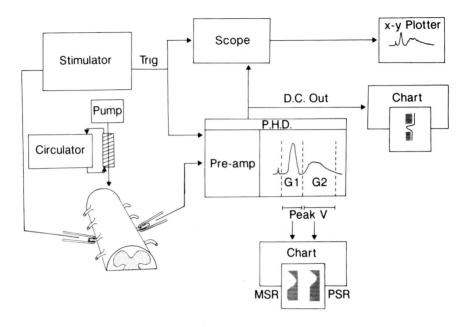

Figure 2. Experimental arrangement for recording reflex responses and DC potential changes. For pharmacological studies, reflex component amplitude is monitored using a peak height detector. As indicated in the block diagram, a pump is used to superfuse the cord with oxygenated saline which passes through a heat exchanger to raise the temperature to 25 °C. Suction electrodes are positioned onto a pair of roots (L_3, L_4, or L_5). The stimulating electrode on the dorsal root is shown on the left. The suction electrode recording from the ventral root of the same segment is shown on the right. The signal is amplified via a pre-amplifier (Pre-amp) and monitored on an oscilloscope (Scope) from the digital store of which a hard copy of the monosynaptic reflex (MSR) and polysynaptic reflex (PSR) can be produced by an X–Y plotter. The amplified signal is also fed to a twin-gated peak height detector (P.H.D.) (see text), which, on receiving a trigger pulse (Trig), detects the peak amplitudes of the MSR and PSR, and feeds long pulses of equivalent voltage to a chart recorder. Slow DC potential changes arising from depolarization of motoneurones are fed to a further chart recorder.

The delays allow selective sampling of the amplified signal during the period of the MSR (5–15 msec after the stimulus) and the PSR (15–100 msec after stimulus), but not during the period of the stimulus artifact. The peak voltages may be fed as 0.5 sec square waves to a potentiometric chart recorder.

The fast flow rate causes rapid exchange of the bath fluid and allows application of drugs via the superfusion system. Reductions in reflex amplitude may be demonstrated with agents blocking non N-methyl-D-aspartic acid (non-NMDA) glutamate receptors, such as kynurenate (10 μM–1 mM) or 6-cyano-7-nitroquinozaline-2,3-dione (CNQX) (1–10 μM). In attempting a quantitative assessment of the blockade, the agent should be superfused until a plateau of depression is achieved at that concentration, and the preparation superfused with drug-free Krebs' solution for long enough to allow full recovery of the response. The transmitters of the descending aminergic projections, e.g. 5-HT or noradrenaline, or agonists at their postsynaptic receptors, may be used to examine the modulation of the reflex responses elicited. Examples of some of the results obtainable can be seen in reference 1.

2.3.2 Motoneurone depolarizations

Suction electrodes may also be used to record the depolarization of the population of motoneurones by amines (e.g. 5-HT, noradrenaline) or amino acids. Such depolarizations spread electrotonically along the axons. The magnitude of the potential recorded via a suction electrode will depend not only on the number of motoneurones depolarized and the degree of depolarization, but also on how close the aperture of the suction pipette is to the cord surface. The closer it is, the less the electrotonic decay of the depolarization of the cell body. The seal of the suction pipette is often not sufficiently tight to prevent considerable instability in the DC record. For this reason, it is often more convenient, especially if a series of depolarizations is to be recorded and concentration–response (CR) curves constructed, if an alternative method of recording from the ventral root is employed.

2.3.3 A simple method for recording motoneurone depolarization

A simple way to do this is to make a Vaseline seal around the ventral root and pass the distal end of the root into a separate recording pool (see *Protocol 3* and *Figure 1B*). The use of calomel electrodes and the stability of the seal allows records of the DC level to be less subject to drift and to change in diffusion potentials than those obtained from suction pipettes.

Protocol 3. Recording DC potential changes from ventral roots

1. Use the bath shown in *Figure 1B*, and mount about 1 cm of hemisected cord, prepared as described in *Protocol 1*. Place the cord cut surface downwards on the Sylgard base. It is often helpful to lay the cord on a strip of tissue (e.g. nappy liner tissue, see reference 3).

2. Obtain two suitable calomel electrodes, e.g. pH reference electrodes (Beckmann). These will last very much longer than silver/silver chloride electrodes. Note that the pools into which these should be dipped (*Figure 1B*) make no direct connection with the tissue.

3. Attach the calomel electrodes via insulated leads to a suitable low drift DC amplifier. Obtain a permanent record of DC level from a potentiometric chart recorder.

Protocol 3. *Continued*

4. Note the substantial amount of spontaneous synaptic activity arising from the cord seen as noise on the chart record.[a]
5. Switch the superfusate (flow 2–3 ml/min) to one containing 1–100 μM 5-HT, or to one containing 1–100 μM noradrenaline for periods of 30 sec, and record the depolarizing responses to 5-HT or noradrenaline elicited. A wash period of 20–30 min will be necessary between applications to avoid tachyphylaxis (for further details see references 3 and 4).
6. Construct a CR curve from three to four responses to different concentrations of 5-HT. Concentrations of 1, 10, and 100 μM 5-HT are suitable; 100 μM 5-HT normally elicits a maximal depolarization. Note that the use of intermediate concentrations will increase the time required to construct a CR curve and may introduce discernible tachyphylaxis.
7. Limit the number of different concentrations of 5-HT if the CR curve is to be repeated in the presence of an antagonist, etc. Wash for 1 h in Krebs' solution before attempting to elicit a second CR curve.
8. A number of antagonists will block 5-HT depolarizations. Allow an adequate time (at least 1 h) for the antagonist to equilibrate with the tissue before attempting to construct a second CR curve.[b]

[a] This can be totally eliminated by adding tetrodotoxin (TTX, 0.1 μM) to the superfusion medium. TTX will also block the reflex response elicited by stimulation of a dorsal root.
[b] Ketanserin, ritanserin, spiperone, mesulgerine, and cyproheptadine are all effective antagonists, but rightward shifts in the CR curves may be accompanied by depression of the maximum (see reference 3).

2.4 Significance of uptake mechanisms in the cord

The sensitivity of neurones in the cord to 5-HT is considerably diminished by neuronal uptake of the amine into nerve terminals. In consequence, the EC_{50} for the depolarizing action of 5-HT is about 20–30 μM when determined by the method of *Protocol 3* (5) or from intracellular recording from motoneurones (6). After blockade of uptake with citalopram (0.1 μM), the EC_{50} is 1.4–3.6 μM. The sensitivity of reflex responses to 5-HT is enhanced by a much greater factor (1). 5-HT depresses the MSR and the PSR with an IC_{50} of around 9 μM. After uptake blackade with citalpram, the IC_{50} for depression of the MSR is 0.03 μM and for depression of the PSR is 0.09 μM.

Our experiments indicate that a certain amount of 5-HT is released spontaneously from the tryptaminergic pathways, because citalopram (0.1–1 μM) itself causes a depression of reflex responses in many preparations. Since this depression is blocked by the 5-HT$_2$ receptor antagonist ketanserin and is elicited by other uptake inhibitors, it is probably due to endogenous 5-HT accumulating and inhibiting the reflex pathways.

2.5 Ipsilateral and contralateral reflexes

If 1–2-day-old rats are used, viable preparations can be made from several segments of whole lumbar spinal cord without hemisection (see reference 7). Contralateral as well as ipsilateral reflexes can be recorded from such a preparation and, by stimulating high-threshold afferents in the dorsal roots, very slow reflex components are elicited which may in part be mediated by substance P. The method is described in *Protocol 4*.

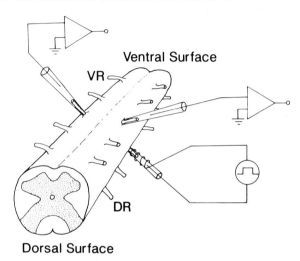

Figure 3. Experimental arrangement for recording ipsilateral and contralateral reflexes from whole lumbar cord of 1–2-day-old rat. The cord is positioned dorsal surface downwards. Suction electrodes are used to stimulate a dorsal root (DR) from segment L_4 or L_5 and to record from the ipsilateral and contralateral ventral roots (VR) of the same segment.

Protocol 4. Recording fast and slow reflex components from whole cord

1. Prepare the cord as in *Protocol 1* but select a 1–2-day-old rat. In the UK, Home Office approval is again required. Omit the stage of hemisection.
2. Use a bath of the type shown in *Figure 1A*. If the preparation needs to be tethered, take care in doing this because of its extreme delicacy. Pin a moist nappy liner strip at either end, but so that the pins do not touch the cord, to tether it or devise an alternative method.
3. Place the cord on its dorsal surface with the ventral surface uppermost. *Figure 3* shows the arrangement of suction electrodes on both ventral and on one dorsal root from the same segment. Hold these electrodes in micromanipulators.

Electrophysiological techniques

Protocol 4. *Continued*

4. Superfuse the cord with oxygenated modified Krebs' solution or aCSF at 4–5 ml/min at 19–21 °C.

5. Use square wave stimulus pulses, 0.1–0.5 msec in duration, and vary the voltage to obtain maximal reponses. If repetitive stimulation is required, use a stimulus frequency of about 0.015 Hz which allows responses to be elicited repeatedly without significant decrement.

6. Examine the fast transients on an oscilloscope. Note that an MSR can be recorded from the ventral root ipsilateral to the dorsal root on which the stimulating electrode is placed. A guide to the viability of the preparation is MSR amplitude. Check that the MSR is of constant amplitude and more than 1 mV.

7. For convenience, use a microprocessor with a signal averaging programme to sample responses, e.g. a CED 1401. Use a laser printer, e.g. Brother HL-8e, to make satisfactory hard copies of the response.

8. Measure the slow components of individual responses from a continuous plot of DC potential on a potentiometric chart recorder (e.g. Bryans (Gould) BS-272).

9. Note that five distinct responses can be recorded from the ipsilateral and contralateral ventral roots (*Figure 4*). In addition to the MSR and PSR recorded from the ipsilateral ventral root, observe the slow response which can be recorded; it has a half-decay of about 8 sec. Note the two main components which are elicited in the contralateral ventral root: a relatively fast reflex and a slow reflex with a half-decay of about 8 sec.[a]

[a] The amplitude of the contralateral reflexes is normally less than 0.5 mV.

The five reflex components are conveniently referred to as MSR, PSR, and ipsilateral slow (IPSI SLOW) for the ipsilateral responses and contralateral fast (CON FAST) and contralateral slow (CON SLOW) for the contralateral responses. MSR, PSR, and CON FAST are elicited by stimulation of lower threshold dorsal root afferents, whereas to elicit maximal IPSI SLOW and CON SLOW responses higher voltages are required because of the involvement of C fibre afferents.

2.5.1 Voltage–activation curves

Voltage–activation curves for the different reflexes can be determined using 0.1 msec and 0.5 msec square wave pulses to the dorsal root, and using voltages between 0 and 100 V. The amplitude of the resulting response can be plotted as a percentage of the maximum amplitude of that component and plotted against the logarithm of the voltage. Better separation of the curves is obtainable using 0.1 msec pulses, providing that maximum slow reflexes can be elicited in the particular preparation using this width of pulse. This is

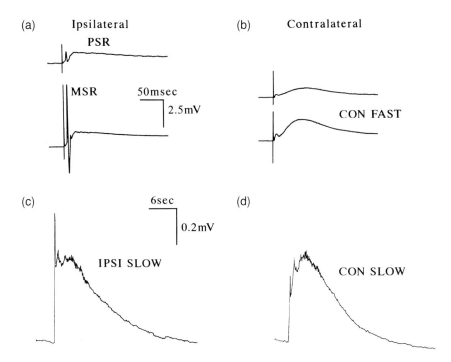

Figure 4. Ipsilateral (a, c) and contralateral reflex responses (b, d) recorded on stimulation of a lumbar dorsal root from the neonate rat spinal cord. (a) A submaximal stimulus (8.5 V, 0.1 msec) evokes a short latency (monosynaptic reflex, MSR) and a longer latency reflex (PSR) in the ipsilateral ventral root (upper trace). The middle trace shows these components elicited by a supramaximal stimulus (20 V, 0.1 msec). (b) The same stimuli evoke a relatively rapid reflex in the contralateral ventral root (CON FAST). In some preparations, a distinct short latency response is detectable (time calibration applies to (a) and (b), voltage calibration is 0.5 mV for (b). (c, d) Slow responses generated ipsilaterally (IPSI SLOW) or contralaterally (CON SLOW) are shown; stimulus 20 V, 0.5 msec; calibrations apply to both. Note that the rapid components are attenuated by the chart recorder.

mainly dependent on the adequacy of the seal made by the stimulating suction pipette. A typical set of results is shown in *Figure 5*. The reflex component of lowest threshold is PSR. With 0.1 msec stimulus pulses, half-activation of this component is achieved with a voltage of about 6.0. Half-activation of MSR and CON FAST is achieved with similar voltages of 8.1 and 7.7, respectively. Half-activation of IPSI SLOW requires a voltage of 10.9 and of CON SLOW a voltage of 16.3 (see reference 8).

2.5.2 Involvement of glutamate receptors

The five reflex responses can be shown to be mediated via glutamate receptors. The selective kainate/AMPA receptor antagonist, CNQX, produces near total blockade of all five components of the reflex responses at

Electrophysiological techniques

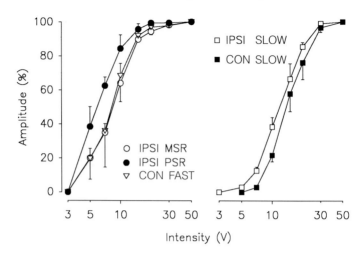

Figure 5. Voltage activation curves for the five components of ipsilateral and contralateral segmental reflexes. Fast reflex components are shown on the left (IPSI MSR, IPSI PSR, and CON FAST) and slow components on the right (IPSI SLOW and CON SLOW). Each point represents relative amplitude expressed as a percentage of the maximal response (ordinate) plotted against stimulus intensity on a logarithmic scale (abscissa). Values are pooled from five preparations. A stimulus duration of 0.1 msec was able to elicit maximal fast and slow responses and allowed better separation of the activation curves.

a concentration of 10 µM. 1 µM CNQX also causes a blockade of more than 50% of MSR, CON FAST, and CON SLOW, while PSR and IPSI SLOW are blocked in amplitude by about 25%.

The NMDA receptor antagonist (($\pm$)-2-amino-5-phosphono-pentanoic acid (APV), has a more selective action on reflex responses. At a concentration of 5 or 20 µM APV, MSR is little affected by the antagonist and PSR is only marginally reduced. The slow components are, however, substantially attenuated in amplitude by APV. IPSI SLOW and CON FAST are reduced to a similar extent by APV (20 µM), while the component most sensitive to APV is CON SLOW.

2.5.3 Modulating actions

Experiments in my laboratory have concentrated on the possible role of descending tryptaminergic pathways in modulating segmental reflex responses. *Protocols 5* and *6* describe some of a range of experiments which examine the actions of:

- 5-HT
- agonists which selectively mimic the actions of 5-HT on a particular subtype of 5-HT receptor
- stimulating the cord more rostrally with bipolar electrodes

Protocol 5. Actions of 5-HT and selective agonists at 5-HT receptor subtypes.

1. Examine the effects of superfusing 5-HT in concentrations ranging from 0.3–50 μM and observe the concentration-related depression of all five reflex components. Note that the slow reflexes, especially CON SLOW, are inhibited more readily than the faster ones.
2. Construct concentration–inhibition curves to determine the IC_{50} values for inhibition of each reflex component. Plot percentage inhibition against the log of the concentration of 5-HT.[a]
3. Compare the IC_{50}s with those in reference 8. Confirm that the IC_{50}s for depression of MSR and PSR are around 7.6–7.8 μM, while those for depression of CON FAST and IPSI SLOW are around 2.6–2.8 μM. The IC_{50} for inhibition of CON SLOW should be around 1.2 μM.
4. Various selective agonists may be tested. Try, for example, the selective 5-HT$_{1A}$ receptor agonist, dipropyl-5-carboxamidotryptamine (DP-5-CT). Confirm that this agent causes a selective inhibition of the MSR. The IC_{50} for inhibition of MSR should be about 90 nM.
5. Examine the effects of superfusing methysergide in concentrations ranging from 1–100 μM. Note that methysergide acts as an agonist is this preparation. Observe that the drug selectively inhibits MSR while having little effect on the other reflex components. The IC_{50} for inhibition of MSR by methysergide should be around 26 nM (see reference 8).

[a] The IC_{50} is the concentration of drug causing 50% inhibition.

The slow reflexes will be recorded on a chart record which will also display any change in motoneurone membrane potential induced by 5-HT or a selective 5-HT receptor agonist. 5-HT at the higher concentrations used will elicit depolarizations of motoneurones both ipsilateral and contralateral to the stimulated root. This is expected, of course, since the 5-HT reaches both sides of the cord. An example of the depolarizing action of DP-5-CT is shown in *Figure 6*. Note the depolarization evoked on both sides of the cord by 0.1 μM DP-5-CT. DP-5-CT also caused some reduction of the slow reflexes in this experiment, particularly of the CON SLOW response.

2.5.4 Stimulation of the cord surface
Electrodes for stimulating the surface of the cord in a more rostral region can be constructed by plaiting together insulated silver wire. Two sizes of wire that have proved suitable are those in which the silver core has a diameter of 0.13 mm and, for finer electrodes, of 0.05 mm (Clark Electromedical). The

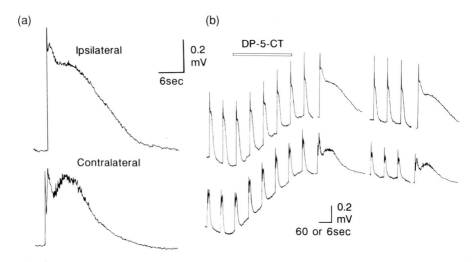

Figure 6. Chart records from ipsilateral and contralateral ventral roots showing depolarization evoked by DP-5-CT and slow reflex responses. The cord is superfused with DP-5-CT (0.1 µM) for 5 min for the period indicated by the white bar. (a) The left-hand traces are control IPSI SLOW and CON SLOW responses. (b) Responses recorded from the ipsilateral root (above) and the contralateral root (below) (note change in calibration). Where the trace is interrupted, slow responses are displayed at a faster chart speed at the peak of the depolarization and on its declining phase.

plaited wire can be secured with Araldite and attached to a holder which will fit into the carriage of a micromanipulator.

If the bare ends of the wire have been cut square, the free end of the stimulating assembly can be lowered perpendicularly onto the cord surface until slight dimpling is visible.

Protocol 6. Stimulation of descending pathways

1. Stimulate the ventral region of the cord between ventral roots T_{11} and T_{12} on one side of the cord to evoke reflex discharges in both ipsilateral and contralateral L_4 ventral roots. Observe that both rapid and slow responses can be generated, depending on the strength of stimulation.
2. Apply a conditioning stimulus to the thoracic cord to produce an inhibition of a subsequent MSR and CON FAST elicited by stimulating a dorsal root.
3. Confirm that this inhibition can, in part, be blocked by the 5-HT_2 receptor antagonist, ketanserin, and thus may be due, in part, to the release of 5-HT from descending tryptaminergic pathways.
4. Employ one of the following chemical strategies to produce a more selective release of 5-HT. Block the re-uptake of 5-HT by a 5-HT neuronal uptake inhibitor such as citalopram.

5. Observe the depression of the MSR caused by citalopram.
6. Superfuse the cord with the 5-HT releaser, parachloroamphetamine (pCA).
7. Observe that, using a concentration of 1 µM pCA, all five reflex responses will be depressed to some extent in the course of the next hour.

It is assumed that the 5-HT released by the ongoing activity in the cord will accumulate to activate 5-HT receptors. After superfusing with 0.1 or 1 µM citalopram (1–2 h), it will be found that segmental reflex responses are depressed, particularly the MSR. This inhibition in the presence of citalopram can be blocked by ketanserin (see Section 2.4).

2.6 Tail-attached spinal cord preparations

A number of laboratories have developed preparations where the lumbar spinal cord is dissected out attached to the tail. This allows stimulation of the skin of the tail with thermal, tactile, or chemical stimuli. Appropriate stimulation causes discharge in, and depolarization of, a lumbar ventral root. Although this technique has recently been introduced into my laboratory, Dray and colleagues at the Sandoz Institute, London, have used it routinely for some time. A brief description of the method is given here, but see also reference 9.

2.6.1 Simple method

A tilted (5°) platform (a 15 × 5 cm Perspex block with large flat area filled with Sylgard) is used to support the isolated cord with tail attached. The spinal cord and tail are separately superfused (spinal cord, 4 ml/min; tail, 8 ml/min) with an oxygenated modified Krebs solution (*Table 1*) at 24 °C. A Vaseline seal and a Vaseline wall built up by laying down ribbons of Vaseline from a syringe can be used to seal the cord from the tail. Activation of peripheral nerve fibres is measured by recording spinal ventral root (L_3, L_4, or L_5) potential change using an extracellular glass micropipette. The ventral root is isolated from the spinal cord by means of a Vaseline well containing modified Krebs solution into which the micropipette is lowered. A segmental reflex response should be elicited from a dorsal root, using a suction electrode, to establish the viability of the preparation. The junction of cord and tail is extremely fragile. That the functional connection between the two remains intact can be established by applying a brief thermal stimulus to the tail surface. This can be done most simply by applying 2 ml of prewarmed Krebs solution (48 °C) to the tail surface. Thermal stimulation excites peripheral thermoreceptors in the skin of the tail and can be applied at intervals of 20–30 min to evoke reproducible excitations of high-threshold sensory endings. Care should be taken with the precise temperature, because we have found that slightly higher temperatures (e.g. 50 °C) sensitize the tail and produce responses of successively increasing amplitude. Capsaicin (0.5 µM) or bradykinin (0.5 µM) can also be used to stimulate sensory nerve endings in the skin of the tail (9).

2.6.2 An improved method

A useful elaboration of this technique is described in reference 10. The spinal cord and the tail are arranged in two adjacent perfusion chambers. That for the tail consists of the disposable plastic tip of a Gilson micropipette and a thin rubber membrane, which covers the open end and fits around the tail. A noxious stimulus can be applied to the tail as a small amount (40–200 µl) of capsaicin solution (0.5–2 µM). This can be injected into the perfusion solution to the tail with a pressure pulse (0.4 kg/cm^2, 0.1–0.5 sec). The pressure pulse is applied to a test tube containing the capsaicin solution, using a nitrogen cylinder and a solenoid valve controlled by an electronic stimulator. To facilitate capsaicin action, the superficial layer of skin throughout the distal four-fifths of the tail should be removed with fine forceps, taking care to remove only the most superficial skin layer. In addition, to enhance the stimulating action of capsaicin, the tail can be perfused with a solution containing 1 µM prostaglandin E_1 for 3 min before applying capsaicin.

The cord lies on a platform and is superfused with an aCSF containing (additionally) 30 mM glucose. Solutions of substance P and other agonists can be applied into the superfusion stream, and a valve can be used to prevent back flow. Concentration-related depolarizations of motoneurones can be recorded from a lumbar ventral root via a glass capillary suction electrode.

2.7 Other uses of the hemisected cord technique

The superfused hemisected cord can also be used to record the dorsal root potential and the dorsal root reflex. Both these responses are elicited in a dorsal root on stimulation of an adjacent dorsal root or of the dorsal columns. Either suction or wick electrodes are suitable for recording. Most of the technical details have already been covered above. The reader is referred for further information to references 11 and 12.

3. Intracellular recording from hemisected cord

3.1 Stimulation of the dorsal root and the cord surface

The technique for preparing the cord is essentially as described under *Protocol 1*. An arrangement for stimulating and recording which we have

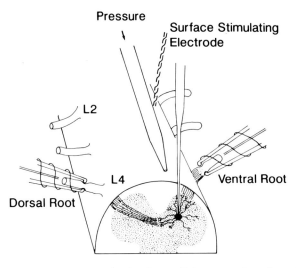

Figure 7. Experimental arrangement for intracellular recording from lumbar motoneurones in the neonate rat hemisected spinal cord. One suction electrode is placed on the segmental dorsal root to elicit orthodromic responses and another is placed on the ventral root to activate the motoneurones antidromically. The cord is continously superfused with oxygenated Krebs solution. A pipette containing an agonist is positioned lateral to the ventral root above the cord and is connected to a pressure ejection system. A bipolar surface-stimulating electrode was fabricated from plaited, insulated silver wire and positioned latero-ventrally on the cord surface two to three segments above the recording site (L_4–L_6).

found suitable is shown in *Figure 7*. For intracellular recording, it is essential to know that the cell impaled is a motoneurone. After impaling a cell in the ventral horn of the cord, this can be confirmed if an antidromic response (preferably a full spike or an initial segment spike) is evoked on stimulation of the appropriate ventral root. The ventral root can be stimulated with a suction electrode (see *Protocol 2*). If no such response is evoked, an interneurone or a motoneurone, whose axon travels to another ventral root, may have been impaled.

Intracellular recordings can be obtained from motoneurones using microelectrodes (20–80 MΩ) filled with either 0.5 M KCl or 2 M potassium acetate. Suitable recording equipment, although not needing to be as elaborate as this, could include an Axoclamp-2A amplifier operated in bridge mode (there are cheaper alternatives). Responses can be stored either on videotape, using a PCM-2 VCR adaptor, or on computer disc using pCLAMP 5.5 (Axon Instruments) for later analysis.

A stimulating suction electrode positioned on the segmental dorsal root (*Protocol 2*) allows orthodromic responses to be evoked in the motoneurone; monosynaptic and polysynaptic excitatory postsynaptic potentials (EPSPs) can be elicited. Possible monosynaptic activation of a motoneurone from a 'descending' pathway can be achieved using a fine bipolar electrode (see Section 2.5.4 or use a commercially available concentric electrode SNE-100 semi-micro, Clark Electromedical) placed on the cord surface. This can be positioned latero-ventrally, two to three segments or more above the recording site. The responses evoked may include those from long-process interneurones whose axons project down the cord. Fast EPSPs and fast inhibitory postsynaptic potentials (IPSPs) are normally elicited from stimulation at this site. In addition, the responses may include slower synaptic responses (slow EPSPs), especially when a short train of stimuli (5–50 Hz, 10–50 stimuli) is used. Slow EPSPs may be mediated in part by release of 5-HT, since they can be reduced in amplitude by blockade of 5-HT$_2$ receptors (13).

3.2 Application of transmitter chemicals

If it is desired to study the action of depolarizing neurotransmitters, etc. on motoneurones, an agent can be applied by superfusion. In our experiments (6), concentration-dependent depolarizations have been evoked by superfusing 5-HT (10–300 μM) for 5 min or by superfusing noradrenaline (1–30 μM) for 5 min. These substances produce similar effects on motoneurones, which include depolarization accompanied by an increase in input resistance, and an increase in excitability to injected current. pCLAMP will control a programme of injected current steps, both in the hyperpolarizing and depolarizing directions, which can be initiated both before and after application of an amine.

To construct full CR curves, it will probably be necessary to pool data from many cells, perhaps using two or three concentrations on each cell. You will be unlikely to be able to construct a full curve from the responses obtained from a single cell. The assessment of blockade of 5-HT responses by an antagonist may require equilibration of the tissue with an antagonist for at least an hour. In these circumstances, a CR curve based on pooled data can be obtained from cells impaled after the incubation period.

Amines or glutamate can also be applied by pressure ejection or by iontophoresis from a pipette positioned as close as possible to the recording site above the cord surface (*Figure 7*).

3.3 Muscle-attached preparations

The advantages of *in vitro* techniques can be enhanced further by arranging more selective stimulation of the afferent input to motoneurones.

In a technique described by Jahr and Yoshioka (14), certain of the hindlimb muscles are dissected free while leaving intact their nerve connections to the sciatic nerve and spinal cord. This allows afferent input from a specific muscle to be stimulated. Further details are given in *Protocol 7*, but see also reference 15.

Protocol 7. Setting up a cord hindlimb muscle preparation

1. Use a 4–10-day-old rat of either sex.
2. Anaesthetize the rat with ether and then remove the vertebral column, pelvis, and right leg. Place in a dissecting dish so that the tissue can be superfused with oxygenated Krebs' solution (*Table 1*) or aCSF at 20 °C.
3. Carry out further dissection under a dissecting microscope at a magnification of 60–160. Following ventral laminectomy, hemisect the spinal cord and section at a low sacral level.
4. Free L_4 and L_5 dorsal and ventral roots, the corresponding dorsal root ganglia and the sciatic nerve from the vertebral column and surrounding tissue.
5. Dissect the sciatic nerve to the lower leg where the common peroneal, sural, and tibial nerves diverge.
6. Some of the individual muscle nerves which emanate from the tibial and common peroneal nerves can be dissected along with the innervated muscles.[a]
7. Tease the muscles apart in order to expose small branches of their nerves for stimulation (see reference 15).
8. Section the cord rostrally at about T_{12}. Pin the hemisected cord mid-sagittal side up, securing it to the Sylgard base of the recording chamber. Maintain the superfusion medium at 25–27 °C.
9. Use a second and larger trough in the recording chamber to hold the muscles. Lead the sciatic nerve through a break in a suitable thin partition between the cord and the muscle chambers. There should be a slot in the partition which can be sealed with Vaseline.
10. Pin out the muscles and nerves in the larger chamber so that the various nerves can be stimulated individually. Use bipolar hook or suction electrodes.
11. Stimulate to evoke monosynaptic and polysynaptic EPSPs in individual motoneurones, also IPSPs.

[a] e.g. the medial and lateral gastrocnemius, soleus and plantaris, posterior tibialis, flexor digitorum longus and flexor hallucis longus, and the peroneus longus, peroneus brevis, extensor digitorum longus, and tibialis anterior. In addition, the sciatic branches that innervate the posterior biceps femoris and semitendinosus muscles, and the plantar nerves can also be dissected free.

Acknowledgements

The author's work is supported by grants from the Wellcome Trust.

References

1. Crick, H. and Wallis, D. I. (1991). *Br. J. Pharmacol.*, **103**, 1769.
2. Courtice, C. J. (1977). *J. Physiol.*, **268**, 1P.
3. Wallis, D. I., Connell, L. A., and Kvaltinova, Z. (1991). *Naunyn-Schmiedeberg's Arch. Pharmacol.*, **343**, 344.
4. Connell, L. A., Majid, A., and Wallis, D. I. (1989). *Neuropharmacology*, **28**, 1399.
5. Connell, L. A. and Wallis, D. I. (1988). *Br. J. Pharmacol.*, **94**, 1101.
6. Elliott, P. and Wallis, D. I. (1992). *Neuroscience*, **47**, 533.
7. Akagi, H., Konishi, S., Otsuka, M., and Yanagisawa, M. (1985). *Br. J. Pharmacol.*, **84**, 663.
8. Wallis, D. I. and Wu, J. (1991). *Br. J. Pharmacol.*, **104**, 331P.
9. Dray, A., Bettaney, J., Forster, P., and Perkins, M. N. (1988). *Br. J. Pharmacol.*, **95**, 1008.
10. Otsuka, M. and Yanagisawa, M. (1988). *J. Physiol.*, **395**, 255.
11. Preston, P. R. and Wallis, D. I. (1980). *J. Neural Transm.*, **48**, 271.
12. Preston, P. R. and Wallis, D. I. (1980). *Gen. Pharmacol.*, **11**, 527.
13. Elliott, P. and Wallis, D. I. (1991). *J. Physiol.*, **438**, 217P.
14. Jahr, C. E. and Yoshioka, K. (1986). *J. Physiol.*, **370**, 515.
15. Walton, K. D. and Navarrete, R. (1991). *J. Physiol.*, **433**, 283.

A1

Addresses of some suppliers

Abbott Laboratories Ltd, Queensborough, Kent ME11 5EL, UK — Chemicals

Amersham International plc, Lincoln Place, Green End, Aylesbury, Bucks HP20 2TP, UK — RNA agents, Chemicals, Radiochemicals

Arnold Horwell Ltd, 73 Maygrove Road, London NW6 2BP, UK — Watchmakers' forceps

Axon Instruments, 1101 Chess Drive, Foster City, CA 94404, USA — Amplifiers, Data acquisition software

Baekon Inc., 4420 Enterprise Street, Fremont, CA 94538, USA — Injectors

BDH Ltd, Broom Road, Poole, Dorset BH12 4NN, UK — Sylgard

Beckman (International), 22 Rue Juste-Olivier, CH-1260 Nyon, Switzerland — Calomel half-cells

Blades Biological, Scarlett's Oast, Cowden, Edenbridge, Kent TN8 7EG, UK — *Xenopus*

Cambridge Electronic Design Ltd, Milton Road, Cambridge CB4 4FE, UK — Signal averager, Stimulators, Pulse train generators

Clark Electromedical Instruments, PO Box 8, Pangbourne, Reading RG8 7HU, UK — Electrode glass, Ag/AgCl pellets & wire, Kopf microelectrode pullers

Dagan Corporation, 2855 Park Avenue, Minneapolis, MN 55407, USA (UK Agents, Clark Electromedical as above) — Voltage-clamp amplifiers

Appendix 1

Digitimer Ltd, 14 Tewin Court, Welwyn Garden City, Herts AL7 1AF, UK — Neurophore system, Pressure ejection & iontophoresis, A/D VCR adaptors

Dow Corning, Box 0994, Midland, MI USA — Sylgard

Drummond Scientific Ltd, PO Box 700, Broomall, PA 19008, USA (Laser Laboratory Systems Ltd, PO Box 166, Sarisbury Green, Southampton SO3 6YZ, UK) — Injectors

GIBCO: Life Technologies Ltd, Unit 4, Cowley Mill Trading Estate, Longbridge Way, Uxbridge UB8 2YG, UK — Culture medium

Goodfellow Metals Ltd, Cambridge Science Park, Cambridge CB4 4DJ, UK — Micromanipulators

Gould Electronics Ltd, Warrington, Cheshire WA3 7BH, UK — Chart recorders, Colorwriter, Oscilloscopes

IBI Ltd, (subsidiary of Eastman Kodak Co.), 36 Clifton Road, Cambridge CB1 4ZR, UK (PO Box 9558, New Haven CT 06535, USA) — Injectors

Imperial Laboratories Ltd, West Portway, Andover, Hants SP10 3LF, UK — Collagen

Intracel Ltd, Unit 4, Station Road, Shepreth, Royston, Herts SG8 6PZ, UK — General electro-physiological equipment, Biological amplifiers and data recorders, Anti-vibration tables

Invitrogene, 3985 B Sorrento Valley Blvd, San Diego, CA 92121, USA — RNA agents

Leitz (Leica), Davy Avenue, Knowl Hill, Milton Keynes MK5 8LB, UK — Microscopes

List-electronic, Plungstaedter Strasse 18–20, D-6100 Darmstadt/Eberstadt, Germany — Patch-clamp amplifiers

Appendix 1

Microinstruments (Oxford) Ltd, 7 Little Clarendon Street, Oxford OX1 2HP, UK — Microscopes, Narishige micromanipulators, microelectrode pullers, microforge, Fibre optic light sources

Nasco Inc., 901 Janesville Avenue, Fort Atkinson, WI 53538, USA — *Xenopus*

Optical Instrument Services (Croydon) Ltd, (Narishige Dealers), 166 Anerley Road, London SE20 8BD, UK — Micromanipulators, Electrode pullers

Promega, 2800 Woods Hollow Road, Madison, WI 53711, USA — RNA agents

Pharmacia, Davy Avenue, Knowl Hill, Milton Keynes MK5 8PH, UK — Percol

Russell Electrodes, Auchtermuchty, Scotland, UK — Calomel half-cells

Schott Glass Ltd, Drummond Road, Astonfields Industrial Estate, Stafford ST16 3EL, UK — Fibre optics

Sigma Chemical Co, Fancy Road, Poole, Dorset BH7 7HN, UK
(PO Box 14508, St Louis, MO 63178, USA) — Chemicals, drugs

Stratagene, 11099 North Torrey Pines Road, La Jolla, CA 92037, USA — RNA agents

Sutter Instrument Company, 40 Leveroni Court, Novato, CA 94949, USA — Electrode puller

Systat Inc., 1800 Sherman Avenue, Evanston, IL 60201, USA — Software

Technical Products International Inc. (Vibratome), 13795 Rider Trail, Suite 104, St Louis, Missouri 63045, USA (UK supplier: Intracel Ltd, Unit 4, Station Road, Royston, Herts SG8 6PZ, UK) — Vibratome

Transducer Laboratories, Guildford, Farnham, Surrey, UK — Amplifiers

Appendix 1

Vector Laboratories, 16 Wulfric Square, Bretton, Peterborough PE3 8RF, UK — Fluorescein-labelled avidin

Watson-Marlow, Falmouth, Cornwall TR11 4RU, UK — Perfusion pumps

Weiss, John & Son Ltd, Wigmore Street, London WI, UK — Scissors

Worthington Biochemical Corporation, Halls Mill Road, Freehold, NJ 07728, USA — Enzymes

Zeiss Ltd, PO Box 78, Woodfield Road, Welwyn Garden City, Herts AL7 1LY, UK — Microscopes

Index

action potentials 59–60, 95, 159, 205 (fig.), 260–1, 262 (fig.)
adrenoceptors 102
AHP 177–8
AHP neurones 160
AlF–Ringer 67–9
amino acids
 excitatory 98–100, 275–6
 inhibitory 100–1
anaesthetic 193–4
artificial cerebrospinal fluid
 composition 172 (table), 194, 195 (table)
 recirculating system 201 (fig.)

Barth's medium 72 (table)
baths 267
 Perspex superfusion 92, 144 (fig.), 174 (fig.), 202 (fig.), 266 (fig.)
biocytin 154, 179, 184
Bradshaw's model of ion movement 252 (fig.), 254
brain slice, bathing solution 92–3
 preparation 90–1
 recording 92–3
 sectioning 90–1
 stimulation 94
burst generation 232–4

calcium
 channel blockers 98
 concentration, intracellular 18
 currents 22–5, 58, 63 (fig.)
 loading 16
 single channel currents 24
camera lucida 184–5
5,6-carboxyfluorescein 153–4
cell adhesion, control of 10
cell-attached recording 43; *see also* whole-cell recording
cell dialysis, time course 18–19
ciliary epithelial cells 35–41
 non-pigmented cells 38–9
 pigmented cells 37–8
 recording by patch-clamp 41–6
CNQX 99–100, 275
collagen 4–5, 10
computer programs (in models of neuronal firing) 46
contralateral spinal reflexes 273–4, 275 (fig.), 276 (fig.)

cord
 dissection 195–8
 excision 197–8

delayed rectifier 13, 59; *see also* I_{DR}
dialysis, intracellular 19
Dogiel type II cells 156, 157 (fig.)
dorsal root
 activity, spontaneous 205 (fig.), 208–9
 ganglion 3
 reflexes, evoked 205 (fig.), 209 (fig.)
 transmitter chemicals application, stimulation 280–3
double chamber technique 30
Drosophila melanogaster 84
 receptors 84
drug action on synaptic transmission 105, 276–8, 282
drug application, electronic timing 261–3
drug comparative potency in micro-iontophoresis 259–60
drug release, experimental measurement 249–54; *see also* microiontophoresis, micro-pipette
 quantification 256–8
 transport number 257–8
Dulbecco's phosphate-buffered saline 37

electrode filling 244–7
electro-osmosis 248
endplate potential amplitudes 109, 110 (fig.)
enteric neurones 143
 classification 160
 impalement 158–9
 intracellular staining 143–63
 properties 159
enzyme treatment for cell isolation 8–9, 35–8, 51–5
EPSP 98–101, 162, 282–3
equilibrium points (in models of neuronal firing) 221–2
Euler's method 224–5
exchange, pipette and cell interior 17
excised inside-out recording 44
excised outside-out recording 44
excitatory amino acids, *see* amino acids
extracellular current artefacts, control 259–60

Faraday's law 249
fetal calf serum 37
fibroblast 55, 56 (fig.)

Index

field potentials 205 (fig.), 210
FitzHugh model 216–18, 220–1, 222, 233
5-HT receptors 27, 272, 276–7, 282
fluctuation of synaptic potentials 123; *see also* models of synaptic potential fluctuation
  electrotonic distortions 121
  latency 119, 120–1
fluorescence immunohistochemistry 151
fluorescence microscopy 151, 184 (fig.)
fluorescent dye-filled microelectrodes 158
fresh/cultured cells, differences 46–7

GABA receptors 83–4, 100
  $GABA_A$ receptors 23, 83–4, 100–1, 177
  $GABA_B$ receptors 100–102
ganglionated plexus 143
ganglion cells
  culture 8–9
  recording 12–16
glutamate receptors 83, 275–6; *see also* Kainate/AMPA receptors, NMDA receptors
goodness-of-fit tests 125–6; *see also* models of synaptic potential fluctuation

Hanks' solution 5, 8
hemisection of spinal cord 199–200
hippocampus 90, 106
Home Office approval 193, 267
horseradish peroxidase 154
hydrostatic efflux 247–8
hyperpolarization-activated inward current 59–60

I_a 25–6
$I_{b,Na}$ 59–61
I_{Dr} (I_K) 25–6, 59
I_f 59
$I_{K(AHP)}$ 26
$I_{K(Ca)}$ 25–6
I_M 26–7
immunohistochemistry, fluorescence 151
inhibitory amino acids 100
input resistance, *see* membrane conductance, membrane resistance
internal (pipette) solution 56–7, 58 (table)
intracellular electrophysiological recording 158–163, 186, 280–3
intracellular markers, choice 152
  staining 156
intracellular pH 18
intracellular solution, composition 17
inward calcium current 58–9
inward rectifier 47
ion channels, 27, 67; *see also* ligand-gated ion channels

ionotropic receptors 98
iontophoresis, mammalian central nervous system 239–63
ipsilateral spinal reflexes 273–4, 275 (fig.), 276 (fig.)
IPSP 100–102, 162, 177, 282–3

junction potentials 17

Kainate/AMPA receptors 98–100, 275–6
KB solution 52, 53 (table), 54
Kolmogorov–Smirnov test 126
Krebs solution 145
  modified, composition 268 (table)

laminin 4–5, 10–11
Langendorff perfusion 53–5
leak currents 15
  subtraction 16
leak resistance 34–5
lens epithelium 30–5
  double chamber 30, 31 (fig.)
ligand-gated ion channels 23, 27
limit cycle (in models of neuronal firing) 222
Locusta migratoria 84
Lucifer Yellow 152–3, 155, 157 (fig.), 178–9
Luminaea stagnalis 84
luminometric assay 79–80

M current (I_M) 26–7
melatonin 82–3
membrane capacitance 13, 45
membrane conductance measurement 103
membrane currents 13
  run-down 20, 61
membrane potential, time course prediction 215–16
membrane resistance 14, 45, 185–6
Mesocricetus auratus 191
metabotropic receptors 98
microelectrode recording 158, 205 (fig.), 207
microiontophoresis 248–9
  experiment design 254–6
  retaining currents 250–3
micropipette
  drug release 247–9
  diffusion 249
  quantification 256–8
  filling with drug solution 244–6
  injection of mRNA 73–5
  manufacture of blanks 240–1, 245 (fig.)
  multibarrelled 240–1, 245
  profile 243
  pulling 242–3

Index

microscope, dissecting 195, 196 (fig.)
models of neuronal firing
   adaptation 229–32
   burst generation 232–4
   differential equation models 224–5
   repetitive firing 226–9
models of synaptic potential fluctuation 123
   confidence intervals 125
   constrained gradient method 125
   E–M algorithm 125, 127
   iterative methods 125
   maximum likelihood estimator 123
   maximum likelihood method 124–6
   Monte-Carlo simulations 125
   method of moments 123
   noise contamination, deconvolution 126
   non-quantal models 129
   quantal models 130
modulating actions 276–7
monosynaptic reflex 269, 270 (fig.), 275 (fig.), 276 (fig.)
motoneurone depolarization 271
   recording 271–2, 266 (fig.), 281 (fig.), 280–3
multi-component postsynaptic potentials, pharmacological potentials 98–103
muscarinic receptors 23
muscle-attached preparations 282–3
myenteric plexus 143
   histochemical detection 149–50
   preparation 146–7
myocytes 55, 56 (fig.)

Na^+/Ca^+ exchange 60
Na^+-K^+-ATPase 37
Na-K^+-pump 60
neuronal firing patterns, mathematical description 215–35
neurones, visualization 183–5
   epifluorescence microscope 184
neurotransmitter
   depolarization 282
   multiple release 101–2
   receptors 27
   release 98
   release site 111
nicotinic currents 23
nicotinic receptors 23, 84
NMDA receptors 98–100, 276
nodose ganglia, finding 5–7
   dissection 6 (fig.)
non-NMDA receptors 101, 270
non-quantal models of transmitter release 129–30
nullclines 220–1
number of transmitter vesicles releasable 115–16
nystatin 20–1, 61–2, 183

ocular epithelia, transport mechanisms 29–47
oocytes 65
osmolarity 18
outside-out patch 23
outward K^+ current 59

parameter, varying effect (in models of neuronal firing) 222–3
patch amplifier 13
patch clamp 41, 76–8
patch pipettes 176–7
   breaking back 39
   filling 40
   fire-polishing 39
   glass 39
   pulling 39
   Sylgard-coating 40
patch recording 41–4
   perforated 183
   technique 20–1
PDF model 128
peak amplitude measurement 119–20
peak height detector 269–70
perforated-patch recording 20, 183; *see also* permeabilized patch technique
perfusion system 201–3
peripheral neurones, membrane currents in short-term culture 3–27
   cells, preparation 4–12
   ganglion cells, recording 12–16
   pipette/cell interior, exchange adequacy 17–24
   solution design 17
permeabilized patch technique 46, 61–2
phosphate-buffered saline 37, 149
photomicrography 151
pipette solutions 56, 177–9
   additions 20
poly-D-lysine 4–5
polysynaptic reflex 269, 270 (fig.), 275 (fig.), 276 (fig.)
population responses 95, 265
postsynaptic potentials 96–101
   blockade 97
   evoking/recording methods 93–7
   excitatory 281
   excitatory/inhibitory 96 (fig.), 99 (fig.), 102 (fig.)
   inhibitory 282
   ionic mechanism 103
potassium currents 13–15, 25–7; *see also* I_a, etc.
presynaptic fibres, stimulation 93–5
presynaptic inhibition 106–7
probability density function 126–37; *see also* fluctuation analysis of synaptic potentials
Procion Yellow M4RX 154

Index

quantal analysis 105
  models 111, 130–6; *see also* models of synaptic potential fluctuation, synaptic potentials
    compound binomial 133–6
    generalized 131
    interpretation 111–16
    simple binomial 135
    size 105
    stationary 132–3
    structural basis 111–16
quantum 105, 113–14

recording chamber 202
  flow rate 202–3
reflex responses 269–71
  fast spinal 273–7
  slow spinal 273–7
release site 111–13
retaining currents 250–4
reversal potential 103–5, 177
root recording 205 (fig.), 206–7

S neurones 160, 162
series resistance 13, 45
Schistocerca gregaria 83–4
signal-to-noise ratio 261
single-channel recording, current rundown 20–1
single-electrode voltage-clamp recording 96–7
sinoatrial node cells, isolation 49–62
  electrophysiological study 56–7
  membrane currents 58–61
  permeabilized patch technique 61–2
  separation 49–55
  shape 55
site-directed mutagenesis, mRNA translation 81–2
solutions
  competing ion effects 249–50
  pH effects 249–50
spider cell 55, 57 (fig.)
spinal cord
  descending pathway, stimulation 278–9
  extracellular recording techniques 265–6, 268–9
  hemisected, intracellular recording 280–3
  in vitro preparation 189–213
    anaesthetic 193–4
    cord, dissection 195–8
    dissection 191–3
    hemisection 199–200
    perfusion system 201–3
    recording 203–10

temperature 210–11
troubleshooting 211–13
vertebral column, removal 194–5
isolated quantitative pharmacology, electrophysiological techniques 265–83
  lumbar hemisected, preparation 267–8
  muscle-attached preparations 282–3
  parts, identification 204–6
  slice preparation 171–2
    whole-cell patch-clamp recording 169–87
    photomicrograph 175 (fig.)
  stimulation of cord surface 277–9
  stimulation of roots 208, 269, 273 (fig.)
  tail-attached preparations 279–80
  transmitter chemical application 277–9, 280–3
  uptake mechanisms 272
state diagram 218–19, 227 (fig.), 228 (fig.)
state space 218–19
step size, choice 225–6
submucous plexus 143
  histochemical detection 149–50
  preparation 148
superior cervical ganglion 3
  culture 8–9
  finding 5–7
    dissection 6 (fig.)
  rat, cultured on laminin 10–11 (fig.)
switching amplifier 12
Sylgard coating 40
sympathetic ganglion cell
  channels 22–7
  theoretical 19
sympathetic preganglionic neurones
  electrophysiology 185, 186 (fig.)
  intracellular/whole-cell recordings, comparison 185, 186 (table)
  whole cell recording 171
synaptic blockade 97, 209 (fig.), 210
synaptic currents 116–23
synaptic inputs, stimulation 93–5
synaptic potentials
  amplitude fluctuation models 123
  automatic baseline correction 117
  electrotonic distortion 121
  excitatory 160–3
  latency fluctuations 120
  noise contamination 118
  recording/measurement 97–107, 116–23
  spontaneous 122
synaptic transmission
  blockade 97–8
  brain slices, pharmacological analysis 89–107
  electrical recording 90–7
  postsynaptic potentials, analysis 97–107
  quantal analysis 109–39
  sites of action of drugs 105–7

292

Index

tangent vector (in models of neuronal firing) 219–20
temperature 209 (fig.), 210
tetrodotoxin 17
time-dependent currents 58–60
time-independent currents 59 (fig.), 60–1
time integral measurement 119–20
Torpedo 81
total membrane capacitance 13, 14 (fig.)
trans-epithelial potential 32–5
trans-epithelial resistance measurement 34–5
triggered firing 227–9, 231 (fig.)
troubleshooting 211–13
two-electrode voltage clamp 76
Tyrode solution 49, 52, 53 (table), 54

unit recording 201 (fig.), 210

vasoactive intestinal peptide 153 (fig.)
ventral roots 198
  activity, spontaneous 209
  DC potential changes, recording 271–2
  potentials, evoked 205 (fig.), 209–10, 269–71, 273–4
vertebral, cleaning 195, 196 (fig.)
vertebral column, removal 194–5
vesicle hypothesis 115
vesicles 115
Vibratome 90–1, 173
voltage-activation curves 274–5
voltage-clamp amplifier 12–13

whole-cell recording 44–6, 57, 97, 174, 179, 169–87
  blind 180, 179–83
  current rundown 20–1
  spinal cord slices 169–87

Xenopus laevis anaesthesia 71
  surgery 70–1
Xenopus laevis oocytes
  electrophysiology 65–84
    complementary screening techniques 79–80
    co-translation of mRNAs 81
    endogenous characteristics 65–9
    expression studies 69–73, 80–4
    G-proteins 66–9
    morphological properties 66
    mRNA injection 75–9
    preparation 72–3, 77 (fig.)
    properties 66–9
    recordings 75–9
    second-messenger systems 66–9
    translation of RNAs following site-directed mutagenesis 81
    phosphatidylinosital-linked receptor expression, luminometric assay 79–80
    Ringer's solution 72 (table)
    specific subunits, expression/characterization 80–1

Zamboni's fixation 150–1